Local Food Systems and Community Economic Development

Local Food Systems and Community Economic Development provides scholarly and practical knowledge on a range of issues often associated with local food system development.

Many people agree that there are unintended consequences associated with the manner in which our food supply chain has evolved. These concerns range in focus from health, to environment, to economic structure, to social justice. But for each argument critical of our current food system, there are to be found strong counter-arguments; the popular press is replete with stories that lean toward taking specific sides in these arguments, often demonizing those on the other side. In this volume, local food scholars strive to be fair, balanced, and as factual as possible in their arguments. This even-handed approach is appropriate as it should foster more sustainable community change, lead us toward a stronger foundation for scholarly inquiry, and ultimately result in more respect and credibility for efforts to better understand the phenomenon of local and regional food system development.

Amid a deepening interest in local food systems as a community economic development strategy, *Local Food Systems and Community Economic Development* will be of great interest to scholars of community development, rural studies, agriculture, food systems, and rural economy. The chapters originally published as a special issue of *Community Development.*

R. David Lamie is a Professor at Clemson University where his work focuses on agricultural and regional development with an emphasis on next generation farmer development, agritourism, regional food systems, and niche market development for specialty crops.

Steven C. Deller is a Professor at the University of Wisconsin–Madison and a Community Development Specialist with the Division of Extension. His research focuses on trying to understand the changing rural economy with an emphasis on identifying strategies that can improve local economic conditions.

Community Development – Current Issues Series

Series Editor: Paul LaChapelle, *Montana State University, USA*

The Community Development Society (CDS) in conjunction with Routledge/Taylor & Francis is pleased to present this series of volumes on current issues in community development. The series is designed to present books organized around special topics or themes, promoting exploration of timely and relevant issues impacting both community development practice and research. Building on a rich history of over 40 years of publishing the journal, *Community Development,* the series will provide reprints of special issues and collections from the journal. Each volume is updated with the editor's introductory chapter, bringing together current applications around the topical theme.

Founded in 1970, the Community Development Society is a professional association serving both researchers and practitioners. CDS actively promotes the continued advancement of the practice and knowledge base of community development. For additional information about CDS, visit www.comm-dev.org.

Community Development and Democratic Practice
Edited by Paul Lachapelle and Michael Rios

Entrepreneurship, Community, and Community Development
Edited by Michael W-P Fortunato and Theodore R. Alter

Collective Impact and Community Development Issues
Edited by Norman Walzer, Liz Weaver and Catherine McGuire

Turning the Tide on Poverty
Edited by Lionel J. Beaulieu and Alice Diebel

Business Retention and Expansion (BRE)
A Practical Approach to Economic Development
Edited by Michael Darger, Alan Barefield and Brent D. Hales

Local Food Systems and Community Economic Development
Edited by R. David Lamie and Steven C. Deller

You can see further details about this series on the Routledge website on www.routledge.com/books/series/CDS/

Local Food Systems and Community Economic Development

Edited by
R. David Lamie and Steven C. Deller

Routledge
Taylor & Francis Group
LONDON AND NEW YORK

First published 2020
by Routledge
2 Park Square, Milton Park, Abingdon, Oxon, OX14 4RN

and by Routledge
52 Vanderbilt Avenue, New York, NY 10017

Routledge is an imprint of the Taylor & Francis Group, an informa business

First issued in paperback 2021

British Library Cataloguing in Publication Data
A catalogue record for this book is available from the British Library

ISBN13: 978-0-367-85976-3 (hbk)
ISBN13: 978-1-03-208184-7 (pbk)

Typeset in Myriad Pro
by Newgen Publishing UK

Publisher's Note
The publisher accepts responsibility for any inconsistencies that may have arisen during the conversion of this book from journal articles to book chapters, namely the inclusion of journal terminology.

Contents

Citation Information vi
Notes on Contributors viii

Introduction: Motivations for a book on local food systems development 1
R. David Lamie and Steven Deller

1 Local foods systems and community economic development 4
Steven C. Deller, David Lamie, and Maureen Stickel

2 Rural wealth creation of intellectual capital from urban local food system initiatives: Developing indicators to assess change 31
Todd M. Schmit, Becca B. R. Jablonski, Jennifer Minner, David Kay, and Libby Christensen

3 Food access, local foods, and community health 49
Steven Deller, Amber Canto, and Laura Brown

4 The impacts of the Farmers Market and Local Food Promotion programs 73
Jeffrey K. O'Hara and Carlos Coleman

5 The Local Agrifood System Sustainability/Resilience Index (SRI): Constructing a data tool applied to counties in the southern United States 89
John J. Green, Jim Worstell, and Caroline Canarios

Index 103

Citation Information

The chapters in this book were originally published in *Community Development*, volume 48, issue 5 (2017). When citing this material, please use the original page numbering for each article, as follows:

Introduction

Motivations for a book on local food systems development
R. David Lamie and Steven Deller
Community Development, volume 48, issue 5 (2017), pp. 609–611

Chapter 1

Local foods systems and community economic development
Steven C. Deller, David Lamie, and Maureen Stickel
Community Development, volume 48, issue 5 (2017), pp. 612–638

Chapter 2

Rural wealth creation of intellectual capital from urban local food system initiatives: Developing indicators to assess change
Todd M. Schmit, Becca B. R. Jablonski, Jennifer Minner, David Kay, and Libby Christensen
Community Development, volume 48, issue 5 (2017), pp. 639–656

Chapter 3

Food access, local foods, and community health
Steven Deller, Amber Canto, and Laura Brown
Community Development, volume 48, issue 5 (2017), pp. 657–680

Chapter 4

The impacts of the Farmers Market and Local Food Promotion programs
Jeffrey K. O'Hara and Carlos Coleman
Community Development, volume 48, issue 5 (2017), pp. 681–896

Chapter 5

The Local Agrifood System Sustainability/Resilience Index (SRI): Constructing a data tool applied to counties in the southern United States
John J. Green, Jim Worstell, and Caroline Canarios
Community Development, volume 48, issue 5 (2017), pp. 697–710

Notes on Contributors

Laura Brown is a Community and Economic Development Educator in the Department of Extension at the University of Connecticut, USA.

Caroline Canarios is a Research Associate and Adjunct Instructor of Sociology at the Center for Population Studies and Minor in Society and Health in the Department of Sociology and Anthropology at the University of Mississippi, USA.

Amber Canto is a FoodWIse State Coordinator for the Family Living Programs at the University of Wisconsin–Extension, Madison, USA.

Libby Christensen is a Postdoctoral Researcher in the Department of Agricultural and Resource Economics at Colorado State University, USA.

Carlos Coleman is an Agricultural Marketing Specialist in the Local Food Research and Development Division of the United States Department of Agriculture, USA.

Steven C. Deller is a Professor at the University of Wisconsin–Madison and a Community Development Specialist with the Division of Extension, USA.

John L. Green is Director at the Center for Population Studies and Minor in Society and Health and Professor of Sociology in the Department of Sociology and Anthropology at the University of Mississippi, USA.

Becca B. R. Jablonski is an Assistant Professor in the Department of Agricultural and Resource Economics at Colorado State University, USA.

David Kay is a Senior Extension Associate with the Community and Regional Development Institute in the Department of Development Sociology at Cornell University, USA.

R. David Lamie is a Professor at Clemson University, USA.

Jennifer Minner is an Associate Professor in the Department of City and Regional Planning at Cornell University, USA.

Jeffrey K. O'Hara is an Agricultural Marketing Specialist in the Local Food Research and Development Division of the United States Department of Agriculture, USA.

Todd M. Schmit is an Associate Professor in the Charles H. Dyson School of Applied Economics and Management at Cornell University, USA.

Maureen Stickel is a Senior Manager at IDinsight, based in Lusaka, Zambia.

Jim Worstell is Coordinator of the Resilience Project, Almyra, Arkansas, USA.

Introduction: Motivations for a book on local food systems development

R. David Lamie and Steven Deller

Increasingly, many believe that food is for much more than simple biological maintenance. Our society is replete with examples of how food symbolizes our value structures, our relationships with each other, and the environment in which we live. We welcome friends, new and old, into our homes to share meals. We often do business over lunch or dinner. We welcome new neighbors with baked goods or vegetables from the garden. Religions base many of their traditions on restricting particular types of food or on full-on feasting. All implore believers to feed the hungry.

Cultural traditions vary greatly around the world, but it would be difficult to find an example that did not involve an element of respect for the way their food is produced, handled, prepared, or consumed. Traditionally, food was consumed in rather close proximity to where it was produced. It was only in recent decades that these traditions were disrupted. This disruption was coincidental with increasing human populations and the advent of refrigeration, transportation infrastructure, petroleum-based production inputs, concentrated and specialized large-scale production technologies, and other ingredients necessary to the development of the long supply chain model of providing food. Add to that the dual income household, information technology, and the concomitant overall busyness of modern life. The resulting disconnects have, arguably, played a role in the disruption of the fabric of the human community. As a result, the last two to three decades have witnessed a steady growth in interest in policies, programs, and practices aimed at ameliorating these conditions, and some are looking to food as an essential ingredient.

Many would argue that there are unintended consequences, primarily anthropomorphic but also more broadly construed, associated with the manner in which our food supply chain has evolved. These concerns range in focus from health, to environment, to economic structure, to social justice. But, for each argument critical of our current food system, there are to be found strong counter-arguments; the popular press is replete with stories that lean toward taking specific sides in these arguments, often demonizing those on the other side.

As editors of this special issue, we sought out local food scholars who strive to be fair, balanced, and as factual as possible in their arguments. For community development practitioners, we think this even-handed approach is appropriate as it should foster more sustainable community change. For community development scholars, this approach should lead us toward a stronger foundation for scholarly inquiry and ultimately more respect and credibility for efforts to better understand the phenomenon of local and regional food system development. As we go to press with this special issue, the Federal Reserve Bank of St. Louis just released a new book titled *Harvesting Opportunity: The Power of Regional Food System Investments to Transform Communities* signaling a deepening interest in local food systems as a community economic development strategy and featuring some of the authors of the articles in this special issue.

The series of articles in this special issue provide scholarly and practical knowledge on a range of issues often associated with local food system development.

The first article by Steven Deller, R. David Lamie, and Maureen Stickel, *Local food systems and community economic development*, provides provoking thoughts and insights into how local food systems are performing from an interdisciplinary perspective with perhaps more weight on the community economics of local food systems. This article seeks to ask tough questions that those engaged in community economic development might demand of local food systems advocates. In their search for grounded theoretical frameworks, rigorous analysis, and hard facts they reveal the many challenges remaining for local food system advocates to make their case while indirectly emphasizing that much more academic and scholarly research needs to be conducted. They argue that much of the early research and policy work on local foods lacks quantitative rigor, but that in the past few years the quality of that work has greatly improved.

The second article by Todd Schmit and Becca Jablonski, *Rural wealth creation of intellectual capital from urban local food system initiatives: Developing indicators to assess change*, reports on interesting research focused on the rural diffusion of knowledge gained by rural farmers selling in urban farmers markets in New York City. It demonstrates how knowledge gained from interactions with an urban customer base might benefit the rural regions where these farmers live and operate their farms. Here, they find that the flow of information through the rural-urban networks forged through local food systems has a positive impact on innovation for all that are involved in the systems.

The next article by Steven Deller, Amber Canto, and Laura Brown, *Food access, local foods, and community health*, provides insights into the relationships between food access, local foods, and community health and the necessary and sufficient conditions for public health and how access to local foods plays into these dynamics. After a detailed review of the food and public health literature, they use statistical analysis to measure patterns between local food access and several measures of public health. While they find strong positive relationships suggesting higher levels of local foods is associated with better health outcomes, they warn that causation has not been established. Is it the case that more local foods result in better health, or is it the case that healthier people create a market for local foods? They suggest that future work in this area should focus on causation and not simply correlation.

This is followed by the article by Jeffrey O'Hara and Carlos Coleman that focuses on two specific funding programs targeted to local food systems development, the Farmers Market Promotion Program (FMPP) and Local Food Promotion Program (LFPP). In this study they review a national survey of farmers market administrators and two case studies of grant recipients and find interesting connections between program success and strong stakeholder support and involvement. These lessons are helping them to shape the future of these programs with the intent of not only strengthening the local food system, but the community.

As community developers realize, not all communities are equally capable of successfully navigating and finding success with federal programs. Those who do successfully compete for these grants likely have invested in supporting a successful grant writer and also likely already have gained momentum with related projects. As agency staff resources are diminished and the paperwork required increases, there is a tendency toward a smaller number of larger grants being provided, meaning a smaller number of community projects will likely be supported. Those communities that are well organized and collaborate within a regional context are more likely to be funded. How this squares with the FMPP and LFPP requirements to prioritize applicants in areas of concentrated poverty remains to be seen.

Finally, the last article in this special collection is by John Green, Jim Worstell, and Caroline Canarios. *The Local Agrifood System Sustainability/Resilience Index (SRI): A research note on constructing a data tool applied to counties in the southern United States* focuses on methodological development. Building from previously conducted case studies and theoretical work, the authors operationalize a resilience framework with county-level data, drawing on individual indicators and a summative index to identify local agrifood system patterns in the southern US. They emphasize indicators of local efforts to create adaptive local agrifood systems with the intent of providing development practitioners and policy-makers with tools to inform decision-making.

We are confident that you will find this special issue of *Community Development* to be a useful resource in your work with communities seeking to advance themselves through local food systems development initiatives.

We would like to thank the authors who took the time to share their wisdom and insights and for being good sports during the editorial process.

Last, but definitely not least, we would like to thank the Southern Rural Development Center for providing the financial support for the publication of this special issue.

Disclosure statement

No potential conflict of interest was reported by the authors.

Funding

This publication was made possible through funding from the Southern Rural Development Center, one of four Regional Rural Development Centers funded through the US Department of Agriculture, National Institute for Food and Agriculture. Any opinions, findings, conclusions or recommendations expressed in the publication are those of the authors and do not necessarily reflect the views of the SRDC or the US Department of Agriculture.

Local foods systems and community economic development

Steven C. Deller, David Lamie, and Maureen Stickel

ABSTRACT

Local food system development is a popular strategy employed by many communities in the pursuit of sustainable and equitable economic growth and development. This often includes a range of economic projects including farmers markets, community supported agriculture enterprises, urban farming/agriculture projects, and food hubs. Sometimes it includes intermediated marketing channels like grocery stores and broad-line distributors interested in providing local foods. Despite growing prominence in the policy arena, many of the presumptions of the local food movement remain largely untested. This review provides critical analysis of current research on local and regional food networks and reveals that much of this research currently lacks strong theoretical grounding and quantitative rigor. As community development practitioners and planners play an important role in food system design, organization, policy, and implementation in an effort to produce community or regional wellbeing, it is important for them to distinguish between objective, research-based information and speculative, advocacy-oriented analysis.

Introduction

Fostering local and regional food systems has become a popular focus of community and economic development strategy employed by communities across the United States, Canada, and Western Europe. Local food systems are typified by small scale, localized production with direct-to-consumer sales though mechanisms such as farmers markets or Community Supported Agriculture (CSA) enterprises and increasingly intermediated sales to local grocery retailers, restaurants, and institutions such as schools, hospitals, and even jails. Producers, especially smaller ones, however, often find it daunting to manage moves into these larger intermediated sales markets. This is troublesome because much of the growth potential in local foods systems is in these latter markets. Thus, a somewhat novel intermediary entity, food hubs, are being developed across the country in regions where there is a sufficient number of smaller farms in reasonable proximity to final markets and a desire to intentionally maintain local supply chains with local products not co-mingled with products from afar (Matson & Thayer, 2016).

The growth in local food systems over the past few decades, both in terms of relative size and interest from a policy perspective at the local level, has been impressive. The dollar value of direct sales from farmers to consumers has grown substantially from about $400 million in 1992 to almost $1.3 billion in 2012, and the number of farmers reporting direct sales increased from about 40,000 to over 140,000 over the same time period (Figure 1). The number of farmers markets grew from 1755 in 1994 to 5274 in 2009 (Martinez et al., 2010) and to 8268 in 2014 (Low et al., 2015), an increase of 371% over 21 years (over 16% average annually), and the rate of growth in demand for farmers markets is likely increasing, with the number of communities wanting to support farmers markets often outstripping the number of farmers able to fill them. The number of CSA enterprises grew from 2 in 1986 to 12,617 in 2012, indicating a truly remarkable rate of growth from this new business model (Martinez et al., 2010). The number of dedicated local food distributors, brokers, and aggregators intentionally serving these local foods marketing channels (food hub) increased by 288% between 2007 and 2014, to a total of 302 (Low et al., 2015). While farm to school programs have multiple objects, such as teaching youth about nutrition and the origins of their food, the latest census of these programs (2011–2012 or 2012–2013 school years) revealed some 4300 schools have some form of program in place, an increase of 430% over levels in 2006 (Low et al., 2015; Ralston & Benson, 2015). Many of these programs are now adapting their model to reach new institutional markets like daycare centers, healthcare facilities, prisons, and corporate commissaries. Many are experimenting with hybrid models that incorporate some elements from food hubs, CSAs, and farm to school approaches, indicating that the future will hold a plentiful array of new models (Woods, Ernst, & Tropp, 2017).

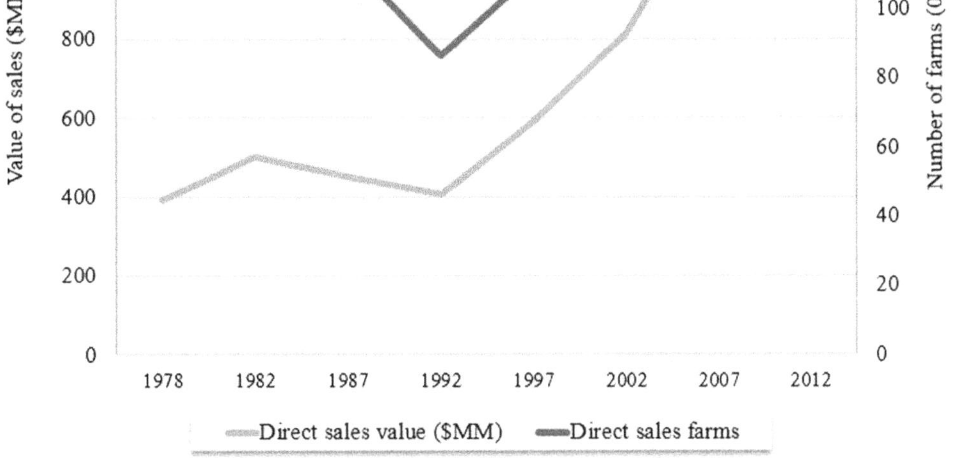

Figure 1. Patterns in direct sales (local foods).

The importance of localized food production and consumption is presented across an increasingly large literature from a variety of perspectives and academic disciplines. Early work on local food systems came mostly from the social sciences including anthropology, rural sociology, and community development (Hendrickson, 2009; Holt & Amilien, 2007). Much of this work focused on the fundamental interest in the ways food is produced, processed, and stored, and the manner in which it is shared within a particular cultural context. This work laid down important foundations of the deeper cultural aspects of food that arguably underlie much of the current interest in local food systems. In addition this work is likely also the reservoir that might feed motivations for changing any given food system to be more congruent with what is truly desired. The literature review, based on a graduate student seminar at UC-Davis, provides a rather comprehensive review and synthesis of these earlier works (Campbell, Carlisle-Cummins, & Feenstra, 2013).

This earlier literature reveals that many authors view the development of a local food industry as at least a partial antidote for a range of societal ills. These societal ills include unhealthy lifestyles and diets, ecologically unsustainable production, withering social capital, food insecurity, and asymmetric economic growth and political power (e.g. Carson, Hamel, Giarrocco, Baylor, & Mathews, 2016). Telligman, Worosz, and Bratcher (2017) note that consumer interest in local and regional foods is connected to socially defined food criteria including improved environmental, social, and economic outcomes. Some, such as Christensen and Phillips (2016), argue that the promotion of local food systems is actually a convergence of community development and economic development efforts and takes on a holistic view of the community. Whether or not one agrees with the arguments advanced by local food advocates, it is clear from the literature that the theoretical frameworks and rigorous analytical testing necessary to support the curative claims is currently lacking and only recently being undertaken in a serious manner. If communities are to pursue policies related to the promotion of local food systems, it is important that these policies be based on a solid research foundation and not merely anecdotal evidence, especially if they are considering large investments of financial, human, or political capital.

In this essay we seek to better understand the strengths and weaknesses of what we know, and do not know, in the local foods research in order to encourage more thoughtful consideration and strategic planning before communities make investments in local food system infrastructure. In conducting this review, we focus on local foods' impact on socio-economic indicators of a community by reviewing literature from the period 2000–2017. Due to the interdisciplinary nature of the range of work on local food systems, literature from nearly all disciplines could apply (e.g. Pinchot, 2014). In our work, we mostly confined our review to articles from economics, sociology, planning, health, and community and rural development. We also confined our investigation of local foods to the developed country context of the United States, Canada, and Europe.[1]

We begin our analysis by discussing how local and regional foods are conceptualized and defined. We then build on our simple descriptive discussion of the growth in local and regional foods and provide some background on the nature of that growth. Next, the local foods literature can be organized in terms of local foods' impact on health, food security, economic development, producer behavior, social capital, and consumer behavior. Within each category, the current literature will be summarized. While the lines between categories often blur, it will be seen that, though much good work has been done, there is a need for more rigorous research. Each phase of our review is undertaken with an eye toward policies, programs, and practices aimed at fostering community and/or economic development.

Defining and describing local foods

A common limitation throughout the literature on local foods is the lack of a standard definition for what constitutes "local" (Cranfield, Henson, & Blandon, 2012). Many authors cite the persistence of subjectivity when it comes to distinguishing between local and nonlocal production and consumption because it is an inherently relative concept. Goodman (2013) explains that food changes based on the context of other ideas, supporting the inability to nail down a specific definition of "local food." McFadden (2015) also highlights the lack of a standard definition. She discusses how a range of attributes such as geography, production practices, business size, and supply chain length play into food being categorized as "local." Thus "local food" can encompass a range of geographic definitions from neighborhood gardens to municipal markets to statewide production (Telligman et al., 2017). This lack of a precise definition of what defines "local" or even "regional" makes any rigorous quantitative analysis of these food systems difficult if not impossible. McCaffrey and Kurland (2014) note that the ambiguity around defining local foods leads to authors and practitioners employing many different definitions of "local" tailored to the primary focus of their work. Thus, the lack of consistent findings about how local or regional food systems impact local communities may be attributable to the lack of an agreed upon definition.

From a practical strategic community action context, however, the precise definition of "local" is less important than movement in the direction of becoming "more local than before," except for decisions about which local governments, organizations, and individuals to involve in their efforts. Community development practitioners have often spent considerable time working with community actors precisely defining terms like "livability," "quality of life," or "local foods," but inevitably find that they must move to action prior to achieving a precise definition. Thus, we can face two competing challenges: a working definition that meets the needs of practical-minded community development practitioners, and a more refined definition that allows us to move our research based understanding of local food systems forward.

Building on McFadden (2015), perhaps a clearer manifestation of "local" appears in the style of production and consumption within the local foods movement. Sundbo's (2013) consumer and producer survey analysis supports this approach despite the ambiguity around the term "local": consumers and producers are demonstrated to have similar ideas as to what constitutes local food. The perception of what defines local foods tends to center on production and purchasing characteristics such as small-scale production and direct to consumer sales mechanisms. These mechanisms include farmers markets, farm stands, and CSAs where customers can connect most directly to a food source. Alternatively, local and regional foods could be generalized as being produced using sustainable production methods and distributed to consumers via shorter supply chains (Bowen & Mutersbaugh, 2014; Garnett, 2013; Hinrichs, 2000; Jarosz, 2008; McFadden, 2015; Sundbo, 2013; Woods, 2012). This focus on shorter supply-chains, both on the input and marketing side, points to the importance of geographical proximity.

A relatively new concept in the local food system arena is the concept of the food value chain (Bloom & Hinrichs, 2011; Diamond et al., 2014). Food value chains are, in essence, supply chains where there is agreement on a particular set of common values that provide a sense of equitable support for each level of the system from producer through final distribution to consumers. The idea is to create value through the transparent delivery of information (the "story") about the individuals involved in the process of producing and delivering the food product to the consumer. A key is emphasizing how each individual is appreciated for the value they add and how they are fairly compensated for their contributions. Consumers who identify with the "story" are thought to be willing to pay more for a product when they know that the proceeds from the sale are distributed equitably.

Farmers markets, CSA enterprises, and other direct to consumer supply mechanisms do not give a full picture of the local and regional food market. As the local foods industry has developed, other entities such as micro-food processers (typically firms with less than five employees) and food hubs have also grown in importance. These entities allow for more coordinated distribution schemes with a wider array of product offerings (Barham et al., 2012; Cleveland, Muller, Tranovick, Mazaroli, & Hinson, 2014; Matson, Sullins, & Cook, 2013; O'Hara, 2017).[2] The growth of these mechanisms points to a role for what Low and Vogel (2011) dub "intermediate sales." These products reach consumers indirectly via sales to entities such as restaurants, hospitals, and jails or to more traditional retail outlets such as supermarkets. Data limitations have constrained the study of these channels, leaving us to case studies of what are often pilot programs. One example would be the study from Maryland that investigates local foods' integration in area school systems (Oberholtzer, Hanson, Brust, Dimitri, & Richman, 2012). Further, as the farms associated with this type of activity tend to be larger and more regionally focused, the lines begin to blur as to what should be counted as "local foods" (Low et al., 2015).

In order to be considered authentically "local," product sold through intermediated market channels requires a "traceability" system that insures the source of local products through the supply chain. This adds cost to the product and most aggregators or wholesalers and distributors have not seen the need since they are generally focused on marketing commodities, not specialized local products. With increased attention to food safety concerns, the regulatory environment is requiring more transparency and traceability, increasing the possibility of these businesses being able to accommodate truly local supply chains as a by-product of investment in traceability systems. Producers who make the effort to brand their products specifically to their farm (estate branding) or in cooperation with other producers (co-branding) can better distinguish themselves in the marketplace, facilitating the sourcing of local foods. In some instances, these branding schemes can become quite formal, including specific rules and regulations governing the geographic location of production, the production methods used, and a whole host of other quality standards intended to protect the integrity of the product and the reputation of the producers. These systems are well developed in Western Europe, but lag behind in the United States, largely because commodity producer organizations have lobbied against them (Barham, Bingen, Lovenworth, Kole, & Mendelson, 2010).

Perhaps a better supported delineation between local and nonlocal rests in the idea of local food networks being an "alternative" to more traditional industrial agriculture and marketing supply chains (Garnett, 2013; Hinrichs, 2000; Jarosz, 2008; Woods, 2012). Lyson (2005) talks in terms of "civic agriculture" which is an umbrella term for community building endeavors in which producers and consumers work together to provide for local food needs. Having smaller scale, localized production and distribution brings consumers and producers closer together thus allowing consumers to better "know their farmer and know their food," a phrase adopted by the United States Department of Agriculture under the Obama Administration.

Characterizing local foods with a quantitative index

The apparent existing and growing interest in local foods development as a policy objective requires that we have more robust methods with which to identify and measure baseline status as well as progress or decline. Given the multi-dimensional nature of local food systems, a host of factors are likely important and should be considered in such an analysis. In this section, we provide a fairly simple approach to help us gain additional insights into the patterns of local foods related activity across the US. We use five separate indicators to develop a "local foods index," realizing that this approach likely only provides partial answers. Another article in this special issue provides an alternative approach to indexing local food systems with a focus on resiliency (Green, Worstell, & Canarios, 2017). Our approach make use of currently available data including the concentration of the number of farmers markets, food hubs, farm to school programs, farms with direct sales to retailers, and community supported farms within each county in the US. These data are drawn from the USDA Food Atlas program that merges several data sources, including the Census of Agriculture, into a common pool database.[3] We define the concentration of the number of each one of these local food activities per ten thousand population. We then use principal components analysis to combine each of these measures into a single index. Principal components analysis utilizes the variance-covariance, or correlation, matrix of the variables of interest to identify common patterns of association. Those patterns can be used to then identify weighting schemes that are used to combine the individual variables into a single index. The goal is that the final index should explain as much of the joint variation across the variables as possible.

The results of this analysis are provided in Table 1. As expected, all five measures move in the same direction (i.e. positively correlated) and higher values of each measure results in a higher value of the overall index. The final index accounts for 53% of the total variation in each measure. Note that CSA farms, farms with direct sales to retailers, and farmers markets contribute the most to the final index, and food hubs and farm to school programs contribute the least. This latter result is likely due to the relative sparseness of food hubs and farm to school programs relative to the other three measures: they are simply less common and hence have lower explanatory power.

A simple mapping of our local foods index (Figure 2(a)) reveals higher levels of local foods related activity in a band from parts of Minnesota through the Great Lakes and into New England along with much of the Pacific coast. There is less activity in parts of the Mountain West, eastern Texas, and parts of Florida. To further explore these patterns, we test for geographic "hot" and "cold" spots using the Getis-Ord method and report those results in Figure 2(b). The Getis-Ord method, along with other variations on the method such as the Local Moran's I, looks for spatial patterns across the units of observations, which in our case are counties. Is there a geographic grouping or clustering of counties that have particularly high

Table 1. Local foods index.

	Weight
Farmers markets	0.459
Food hubs	0.361
Farm to school programs	0.336
Farms with direct sales to retailers	0.519
Community supported Agricultural farms (CSA)	0.525
Explanatory power	0.540

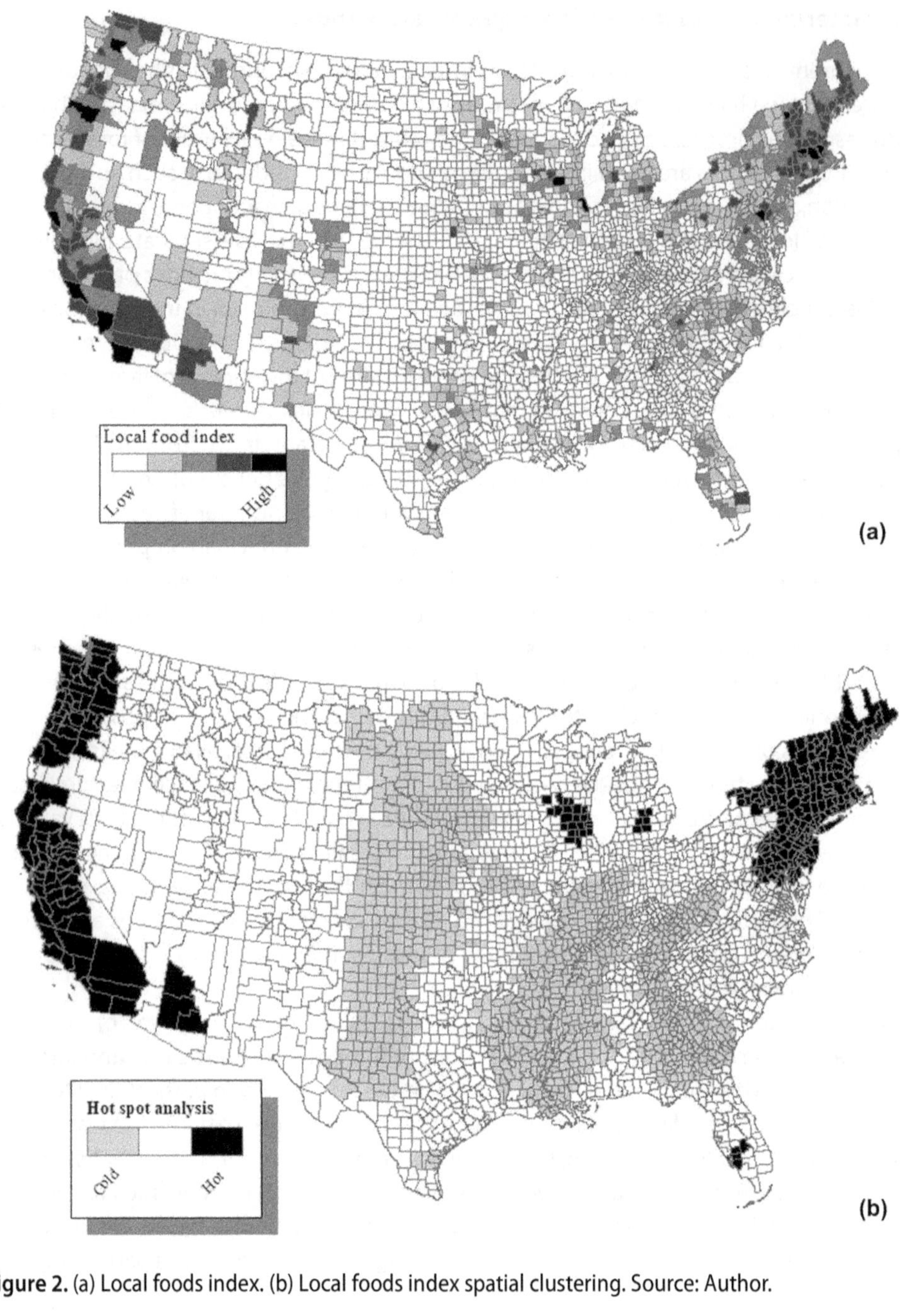

Figure 2. (a) Local foods index. (b) Local foods index spatial clustering. Source: Author.

concentrations (hot spots) or low concentrations (cold spots) of local foods activity, measured by our index? Or is the geographic pattern simply too random for any patterns to be identified? If those concentrations are higher (or lower) than one might expect (the average across all counties given a certain variance around that average), we can draw statistical inferences about those spatial clusters. One way to think about these is if a particular observation is above (hot spot) or below (cold spot) the average at or above the 95% level of confidence, then there are spatial patterns to those statistically significant high and low spots.

Here the pattern from Figure 2(a) becomes clearer: there are "hot" spots of high local foods activity in New England and the northern Atlantic states, southern Wisconsin, south central Michigan, a small part of Florida, and nearly all of the Pacific coast. What is perhaps more interesting is the "cold" spots associated with much of the central Great Plains, the Mississippi River region, and northern Florida north through the western Appalachian region. While low population densities largely explain the "cold" spots in the Great Plains, the lack of significant activity in large parts of the southern US is somewhat unexpected. Based on this analysis, much of the local foods activity is clustered in the northeastern parts of the US, the Pacific coast, and a few pockets in the upper Midwest.

Whether local foods is a geographic concept, an anti-industrial expression, or some mixture of both, what is clear is that local foods has gained much momentum over the past two decades, and this has resulted in documentation of this rapid growth and change (Christy, Landman, Nowatschin, & Blay-Palmer, 2013; Knezevic, Landman, & Blay-Palmer, 2013; Low & Vogel, 2011; Martinez et al., 2010; Nguyen, Wysocki, Treadwell, Farnsworth, & Clark, 2008; Renting, Marsden, & Banks, 2003). This popularity seems coupled with deep-rooted assumptions about what local foods can do for a neighborhood, a community, or an economy. Assertions about local foods include that local food activity helps local economies grow by keeping dollars from leaving, that local foods means fewer middlemen and thus more profit for the farmer, and that local food outlets connect consumers to their food source thus encouraging healthier lifestyles and broader food awareness. In subsequent sections, we develop and explain these assumptions and how they are currently being assessed and tested.

Local foods and consumer demand

While local foods may not be a concretely defined concept, it is clear that there is growing demand for locally grown and processed food products (National Grocers Association, 2015; Low et al., 2015). To gain insights into the growth in demand for local foods, a significant portion of the literature focuses on profiling the typical local foods consumer. Much of this consumer-focused literature is focused on identifying the premium consumers are willing to pay for "local" foods versus more traditionally sourced foods (e.g. Carpio & Isengildina-Massa, 2009; McFadden, Bond, & Bond, 2008; Zepeda & Nie, 2012).

The research offers some conflicting stories possibly due to the regional (e.g. studies within New England and the coastal Pacific region compared to the Mississippi River region) and survey-based nature of the analyses, or it may stem from the lack of consistent definitions. McFadden et al.'s (2008) analysis found that consumers are willing to pay a premium for local foods if consumers believe it will help support local farmers and communities. There exists a strong bias that the local food movement remains prohibitively expensive and thus precludes poorer clientele from participation. Zepeda and Nie's (2012) analysis supports this, indicating that income is the most constraining factor for local food consumption for poorer population segments. If local foods are perceived as primarily "fancy" or "superior" goods, then this make sense. One might expect a different result, however, if local "normal" foods (staple goods) are considered since they are not inherently "fancy." Sadler, Gilliland, and Arku's (2013) work demonstrates that it is not just high-end customers participating in area farmers markets, but that a wider socioeconomic array of clients chooses to buy from local food vendors. While not all local food consumers are wealthy, Willis, Carpio, Boys, and Young's (2013) work does show that willingness to pay increases with income. This is a natural reflection of the ability to pay more as income increases. The dominant theme seems to be that there is a key distinction between consumers not only having the *willingness* to pay more but also the *ability*.

From a producer's perspective, securing space at a market that caters to well heeled, well informed, "foodie" customers is desirable, especially if the producers are capable of delivering high quality specialized food products. Location of direct markets with easy access to these populations is key to their success, so long as market fees and other access costs are reasonable. There are also intentional efforts to reach less wealthy (even lower income) customers with locally sourced fresh food products (mostly fruits and vegetables) via food banks/pantries and "corner store" programs. These programs are typically funded through health organizations and agencies (e.g. the Centers for Disease Control).

Lessons from a failed CSA initiative in a low-income area indicate that whomever the clientele, it is important that local characteristics be taken into consideration when evaluating the viability of markets for local foods (Kato, 2013). Another case study of a failed food hub in Kentucky underscores the importance of careful planning execution of food systems projects such that they do not try to become "all things to all people," but rather that they must pay close attention to the financial bottom line, regardless of the social good that they produce if they are to be sustainable (Brislen, Woods, Meyer, & Routt, 2015). Too often proponents of local food systems proceed on the "build it and they will come" mentality and overlook local preferences. Beyond income level, there appears to be several other key factors that correlate with local food purchases. For example, women appear more willing to pay a premium for local food products than men (Willis et al., 2013). Having young children also increases the odds of being willing to pay for local food (Zepeda & Nie, 2012). More important to local food participation, though, seems to be predisposition to certain hobbies and activities. Zepeda and Nie (2012) find certain behavioral attributes, such as cooking often, having more leisure time, belonging to a health club, or self-identifying as adventurous to be the best predictors of local food preferences. Incorporating these consumer preferences when making venue and product selection decisions would benefit local food vendors, though it is not clear to what extent this is being done.

Many of the characteristics associated with the purchases of local foods are also associated with purchase of organic foods. In fact, there is overlap and confusion between organic and local production. Adams and Salois (2010) argue that the rise of the local food movement coincides directly with the USDA's publication of organic food standards. As they argue, the publication of the standards allowed large, industrialized producers to enter the organic food industry by giving formal definition to what can be marketed as "organic." This event, at least in part, spurred the local food movement to grow as a counter to this change in federal policy. Demand determinants associated with "local" likely involve a preference for smaller farms (small is beautiful) and a direct, face-to-face, intimate connection with the producer. There is a presumed assurance of quality, healthy, and safe food in these relationships and that the proceeds will go to deserving parties.

Local production relates only to the place of production, whereas organic production references the style of production. Adalja, Hanson, Towe, and Tselepidakis's (2015) survey of consumers demonstrates that consumers substitute between means of production and location when making consumption decisions. This may be due to the fact that consumers do not necessarily know what is produced locally and when it would be available to them (Chamberlain, Kelley, & Hyde, 2013). In general, though, it appears that consumers value quality over production methodology (Wirth, Stanton, & Wiley, 2011) and that it is societal preference, not price, the dominates food choices with regards to local food (Barnett, Clarke, Cloke, & Malpass, 2005). This latter idea can perhaps be best understood in the context of civic agriculture, a nonmarket style of community organization in which producers and consumers work together to provide for local food needs (Lyson, 2005). Civic agriculture can take many forms depending on the community, but it is always more focused on promoting cooperation and strengthening communities than it is on maximizing profit (Lyson, 2005). The value chain concept discussed earlier also applies in this context.

What appears to actually draw in consumers to local food outlets is a range of normative tenets that consumers associate with local food production (Sundbo, 2013). Some key tenets include the notion that local foods are of higher quality and superior nutritional value, and that the production of local food products is better for the environment (Martinez et al., 2010). A fairly new concept, not highly researched, is that of "nutritional density." The argument goes that long supply chains, enabled by cheap refrigerated transportation, required the development of fresh fruits and vegetables that could be harvested, stored, and shipped easily and without much product loss. Breeding programs focused on product form with much less attention on flavor and nutritional levels. More fragile, often heirloom, products are often proven to be higher in nutritional content, but they do not store or ship well, making them more appropriate for local markets.[4] Thus, robust local food systems characterized by short supply chains could potentially deliver the more nutritionally dense local products, whereas long supply chains cannot. In order for this to occur, consumers would need to understand the nutritional levels in products like specific fruits and vegetables. Currently, these products do not carry nutrition labels. On a related note, nutritionists concede that fruits and vegetables harvested at their peak and flash frozen will maintain their inherent nutritional value even as they enter long supply chains. This presents opportunity for those producing fragile nutritionally dense products to reach consumers at a distance.

Consumers in the southeastern US cite concerns for food safety as a motivating factor in local food purchase (Maples, Morgan, Interis, & Harri, 2013). Knight's (2013) survey of consumers demonstrates that belief in the societal benefits of local foods mattered more to consumers than did cost. This is akin to the "public goods" aspect discussed by McFadden et al. (2008). Consumers trust local food sources and producers over more anonymous distant large-scale producers (Zagata & Lostak, 2012). In a study of Lexington, Kentucky, Woods, Rossib, and Davis (2017) found the farmers markets, the Kentucky Proud label, and the farm-to-restaurant movement as the most identifiable characteristics associated with local foods systems while availability through retail outlets and farm to school programs were not important. A study of the Certified South Carolina Grown program also showed positive results for their statewide branding program (Carpio & Isengildina-Massa, 2009).

It appears there is no static profile for the consumer of local foods as it can change across location, product, and time. Further, there seem to be conflicting ideas as to why buying local foods is important to consumers: is it preferences for quality, perceived environmentally sustainable production methods, a desire to support local farmers and/or the community, or perceptions of health benefits? Or, is it that shopping at farmers markets or being a member of a CSA is valued as a form of social engagement? It is likely some combination of all of these factors. Customers demand "authenticity" and search for an "experience" when they participate in farmers markets (Wittman, Beckie, & Hergesheimer, 2012). What makes participation "authentic" or an "experience" though is uncertain. A clearer understanding of the consumer of local foods is necessary. Once consumer demand is better understood, it remains to be seen which producers have the savvy to deliver to these expectations. This lack of understanding of local food consumers might be due to the fact that there is yet to be evidence beyond anecdotal experiences and preliminary findings supporting the specific benefits associated with local foods. Shedding further light on the evidence might help us understand what authenticity can and should look like.

Local foods and economic development

A common reason cited for increased attention to developing local food systems is positive growth effects on local and regional economies (McFadden et al., 2016). Additionally, growth of local foods systems is said to create not just growth, but more equitable, stable, and sustainable growth. Even further, these benefits supposedly accrue not just to farmers, but to the broader community as a whole (Lamie et al., 2014). In this section, we explore how the literature has answered questions surrounding local foods' impact on local and regional economic growth. We will pay particular attention to the differences in the literature between economic impact assessment, economic growth, and economic development that, unfortunately, are all commonly used interchangeably (McFadden et al., 2016). It will become clear that there are still gaps in the current literature in terms of justifying the promotion of local foods as a tool for economic development.

A large portion of the current literature as it relates to local foods' relationship to the local economy attempts to capture current industry activity. Some articles describe the role of local foods in terms of a broader multi-dimensional development policy (McFadden et al., 2016). This could be simply providing a broad overview of current activity (DeWeerdt, 2009; Martinez et al., 2010; Storey, 2009) or a look at how the industry has developed across time (Eaton, 2008). Other articles estimate the size and impact of local food entities such as CSAs, farmers markets, and food hubs using traditional economic impact assessment such as input-output multipliers, case studies, and interviews (Barham et al., 2012; Goodman, 2004; Henneberry, Whitacre, & Agustini, 2009; Hughes, Brown, Miller, & McConnell, 2008; O'Hara & Pirog, 2016; Sadler, Clark, & Gilliland, 2013; Sadler, Gilliland, & Arku, 2013).[5]

Important to note is that not all studies successfully account for the opportunity costs associated with switching toward localized production (Jablonski, Schmit, & Kay, 2016). Opportunity costs can be incurred by a range of actors and through a variety of channels. The general idea is that expenditures on local foods are likely diverted from expenditures elsewhere in the economy. Perhaps the most common example of this in the local foods literature is with regards to how economic multipliers employed in estimating economic impacts use gross changes as opposed to net changes. For example, consider the consumer who decides to spend $100 on local foods at a neighborhood farmers market (or CSA, or other local food retail outlet). For a given fixed food budget this spending represents a shift in activity away from traditional retailers (e.g. grocery stores) – or even other non-food purchases – toward the neighborhood farmers markets. Most multiplier based studies, though, consider this $100 in spending as new activity, not a transfer from one type of spending to another. This likely results in an overstatement of the impacts. To more accurately assess the impact, the net change in spending should be examined, not the gross spending. In some cases, however, overall expenditures may increase due to consumers actually increasing their food budget, but likely at the expense of savings or other spending.

Hughes et al. (2008) show how accounting for these costs reduced the overall size of the estimated economic impacts of farmers markets, as expected. Important to note though, is that this is not the only type of opportunity cost that can occur. Looking beyond the community, there are additional opportunity costs to consider, such as how the switch toward local consumption affects the prior (nonlocal) suppliers. Ballingall and Winchester (2010) use the local foods movement in Europe to demonstrate how strong local food movements can negatively affective low-income countries that are highly dependent on food exports. For example, shifting to locally grown vegetables in the US can have a negative impact on vegetable producers in, say, Mexico.

Within the context of describing the potential economic implications of local foods, several authors focus on projected future growth and estimate the size of market potential for various local foods entities. Examples include estimates of production from farmers diverting land use from an export-oriented commodity toward vegetable and fruit production targeting local markets (Swenson, 2011), projections of urban food production using available green space (Kremer & DeLiberty, 2011), and case studies of group purchasing schemes (Little, Maye, & Ilbery, 2010). Others work to evaluate the feasibility of making such a switch. One example would be Cleveland investigating the feasibility of becoming self-reliant in the production of fresh produce (Grewal & Grewal, 2012).

Articles advocating for the growth of the local food industry through a particular channel or mechanism are common in the local foods literature. Examples of this include growth via the refinement of food hubs, further proliferation of CSAs, enhancing the popularity of farmers markets, or creation of local agricultural clusters (Brasier et al., 2007; Brown & Miller, 2008; Matson & Thayer, 2016; Matson et al., 2013; Stagl, 2002; Taylor & Miller, 2010). In all of these analyses, questions of opportunity costs are rarely addressed. While it may be clear that growing vegetables with land can generate sales, is this the *best* use of the land? Are there other, more profitable ventures that could be pursued with this land? Would the local or regional economy grow even faster if resources were devoted to supporting other local enterprises?

Because the aforementioned forecasting and estimation articles fail to address opportunity costs in their analyses, they paint a somewhat disproportionately rosy picture of the growth prospects for the local foods industry. Perhaps for this reason there exists a strong sentiment of doubt and skepticism in the literature. To begin, there is doubt surrounding the existence of positive growth due to local foods activity. It must be made clear that most of the cited articles above remain quite speculative. This is because these articles tend toward focusing on a specific region (i.e. state, city, or farmers market), and therefore the results may not be generalizable. In fact, work modeling income growth as an outcome of local foods activity suggests a weakly positive relationship (at best) between growth and local foods activity (Deller, Brown, Haines, & Fortenbery, 2014). Brown, Goetz, Ahearn, and Laing (2014) use statistical analysis to model US county economic growth and find that a $1 increase in farm sales led to an annualized increase of $0.04 in county personal income. They conclude that, with few exceptions, local foods did not make a significant contribution to economic growth during the study period.

Even assuming there are positive economic outcomes, many authors express uncertainty with regard to the scalability and replicability of many of the local foods models (Albrecht et al., 2013; Blay-Palmer, Landman, Knezevic, & Hayhurst, 2013; Jarosz, 2008; Thilmany et al., 2013; Wittman et al., 2012). This is partially due to the fact that the local infrastructure and the regulatory framework necessary to grow the industry is either not in place or not currently designed to support local foods (Donald & Blay-Palmer, 2006). The Food Safety Modernization Act of 2010 should help address some of the regulatory concerns, particularly for small-scale processors of local foods. Further, since local foods activity relies on a very specific clientele, if these consumers' preferences change, it is unclear if the entire local foods system would be undermined. Several authors discuss successful models, initiatives, and ideas that could and should be replicated in other communities, yet none identified a successful replication. What may be successful in one community will not necessarily be successful in another.

Across the board, the analytical rigor with regard to analyzing local food's impact on economic growth remains a concern. This is in large part due to the limited availability and inconsistency of data forcing researchers to focus on local or regional estimates, case studies, and survey/interview-based analysis in order to understand the effects of local foods activity. Sharp, Clark, Davis, Smith, and McCutcheon (2011) develop three diagnostic tools local communities can use to understand how local food development is affecting their economy. Boys and Hughes (2013) also suggest the application of different regional economic modeling approaches to better understand the impacts of local food movements. McFadden and colleagues (2016) lay out an approach for thinking about local foods systems within a community economic framework, but their work focuses on economic impacts as opposed to economic growth.

Yet, results of these efforts remain forthcoming. As the industry continues to mature and grow, further analysis is needed to help determine best practices and their effects on local economies. It remains to be seen if the local foods market is sufficiently thick for farmers and other businesses in the supply chain to generate sufficient revenues to be profitable at a reasonable level, or even if the market is of a sufficient size to truly impact the overall economy. The next section explores a particular aspect of local foods' interaction with the economy, and this is how local foods activity affects the producers.

Local foods and impact on producers

One key facet of the economic development argument in support of local foods is how buying and consuming local foods "helps the farmers." The general idea is that purchase of locally grown and processed food shortens and simplifies the supply chain and thus the benefits of a local foods sale accrue directly to the local producers instead of to geographically distant middlemen (i.e. wholesalers and distributors).

Many authors have polled farmers active in local foods in an attempt to establish a general profile for a local food producer including producers' motivations for involvement with local foods and identifying indicators of success (Ahearn & Sterns, 2013; Hughes et al., 2008; Izumi, Wright, & Hamm, 2010; Milestad, Bartel-Kratochvil, Leitner, & Axmann, 2010; Wittman et al., 2012). A common finding is that farmer participation in local foods is often understood to be an expression of personal values. More specifically, producers favor local foods channels because they are a means of countering industrial production (Milestad et al., 2010). Many farmers place a premium on activities that involve personal interaction with consumers and will avoid scaling up production to the point where they cannot directly engage customers. In this way, consumer and producer engagement with local foods can be seen as quite similar. Again, the value chain concept discussed earlier applies.

This is not to say that local foods producers are not business-minded. At the very least, local food activities must provide some reasonable return to the farmers in order for these farms to remain in operation. Further, while non-monetary reasons for establishing and participating in a local foods system are widely supported, not all producers engage in local foods as an expression of values. There are local producers who are trying to capture the higher profits possible through the high-end production of organically grown fruit and vegetables (Eaton, 2008; Peterson, Barkley, Chacón-Cascante, & Kastens, 2012). These farmers see market potential and enter that market with the incentive to make profits. This view suggests that local producers are simply responding to a growing demand in the marketplace (Wittman et al., 2012).

Whatever the reasons for participation in local foods activity, though, it is becoming apparent that local food involvement is not guaranteed to be a profitable business strategy for local producers. Individual level producer surveys revealed that most farms engaged in local food production are not profitable unless farmers exclude the value of their own time (Brown & Miller, 2008). Farms that do not allow farmers to earn returns are not economically viable in the long run. Additionally, many farms must supplement direct, local sales with sales to nonlocal, large-scale customers (Milestad et al., 2010).

Many organizations and municipalities are recognizing this deficit and have created programs to support farmers who are predominately supplying local markets. These include not only educational programs around production techniques, marketing and business planning, but also forums to provide networking opportunities (Hightower & Brennan, 2013). Perhaps more important is the question of whether farmers who engage with local foods markets are better off than those who do not.

In sum, one of the predominate arguments for supporting local foods systems is that it retains more of the food dollars in the local community and allows more of the food dollars to flow to the farmer. This in turn results in positive economic impacts and promotes economic growth and development. Unfortunately, outside of a handful of successful local foods producers, it is not clear from the available literature if entering the local foods market is a viable business venture for farmers. What we can conclude based on the literature is that if the goal of the local food movement is to allow for the enhancement of quality of life for the small farmer, which entails ensuring that the business enterprise is viable, it appears more than simply promoting farmers markets and CSAs needs to be done.

Local foods and public health

As we will illustrate in the following section, the argument for local foods with the most quantitative research foundation is in the arena of health. The general claim is that interaction with the local food industry improves the health of a community through increased health awareness and access to healthy food. In our discussion, health is defined to include both bodily health as well as healthy food habits. A part of healthy eating is access to healthy food. Food access, or food security, is often discussed in the context of food deserts. This is a newly growing focus of the local foods research, and our review here is meant as a broad overview of the topic, discussed in much more detail in Deller, Canto, and Brown (2017).

As noted above, the definition of local foods is not always clear and concise, and this is evident within the local foods literature that focuses on health related issues. Some focus on access to farmers markets, while others focus on access to community gardens or farm to school programs. The presence of community gardens, for example, is shown to promote healthy lifestyles in general (Armstrong, 2000; Teig et al., 2009). School programs that involve farm visits, menu selection, and cooking lessons featuring local ingredients demonstrate positive impacts on students' nutrition (Berlin, Norris, Kolodinsky, & Nelson, 2013; Ruge & Mikkelsen, 2013). Individual surveys suggest that positive attitudes toward alternative food production are positively correlated with healthier dietary habits (Pelletier, Laska, Neumark-Sztainer, & Story, 2013). In a study of Kentucky farmers market patrons, Webber, Stephenson, Mayes, and Stephenson (2013) demonstrate that patrons have more knowledge of nutrition and consume fruits and vegetables at a level above statewide averages. In a study of farmers markets in North Carolina, Carson and her colleagues (2016) found similar results, in addition to changes in eating habits to include more fruits and vegetables and exploring foods that patrons reported they would not normally consume.

The above articles all reference specific markets, schools, or communities, yet the positive association between local foods and health holds even when examining this relationship at more general levels (Bimbo, Viscecchia, & Nardone, 2012). In fact, empirical evidence supported by advanced statistical modeling demonstrates that increased local food prevalence has positive outcomes for individual weight and health metrics (Berning, 2012). Moreover, in areas where incidence of local foods is low, the impact on health outcomes is shown to be particularly pronounced (Deller, Brown, & Canto, 2014). These findings support a broader role for local foods in terms of public health policy (Salois, 2012). One specific idea is furthered by Garnett (2013), who argues that the best way to change the food system and further improve public health is to incentivize improved nutrition.

As discussed in detail in Deller, Canto, and Brown (2017), the key limitation of these findings is that it is not clear the direction of causality. Do local foods systems encourage people to be healthier? Or are healthier people more likely to participate in local foods' markets? Many of the studies that take on a tone of advocacy, such as Webber and colleagues (2013), suggesting that access to local foods causes or results in the improved health outcomes. But this conclusion on causation is purely speculative: it could very well be that local foods are emerging because of demands on the part of consumers who are already health conscious. Even so, access to local foods markets may prove to be an important amenity to help attract and retain healthy residents. If good health is correlated with age, income, education levels, or other desirable population traits, then the direction of causality might actually be circular, and we might be witnessing a virtuous cycle of causality. Alternatively, this circular causation could result in a snowballing effect where healthier people create a greater demand for local foods that then reinforces healthier lifestyles.

Local foods and food security

A more recent shift within the research on local foods has been toward food security. The argument is that many places exist where there is limited to no access to nutritious food, particularly fresh food: food deserts. If food desert areas were to experience the introduction of food outlets, particularly healthy food outlets, community health indicators for the area would improve. Further, since this disparity in access is particularly pronounced when looking at low-income neighborhoods (Bader, Purciel, Yousefzaseh, & Neckerman, 2010; Walker, Keane, & Burke, 2010), improvements in community health might generate secondary effects on other important indicators such as income.

At face value, the development of local food infrastructure seems a logical solution to provide these food desert areas with healthy food access. As previously noted, consumer surveys suggest that it is the lack of access to local food outlets that drives nonparticipation by low-income groups in local foods activity (Zepeda & Nie, 2012). Thus, many authors suggest focusing on the effects local foods could have on these areas (Dolstad, Woodward, Green, & Mcspirit, 2016; Sadler, Clark, & Gilliland, 2013; Woods, 2012). Further, introduction of a local food outlet in a food desert area lends itself to a before-and-after impact analysis to gauge the true effects local foods has on these areas (Donald, 2013). A review of the literature on the impact of leveraging federal food and nutrition programs provides insights into their effectiveness, including SNAP/EBT strategies at farmers markets in food deserts (Byker, Misyak, Shanks, & Serrano, 2013).

Some evidence does already exist supporting this concept, though it is limited. Analysis of specific interventions shows the introduction of local foods as having a positive impact on food desert areas. Larsen and Gilliland's (2009) work demonstrates that the price of a bundle of healthy foods decreased at area supermarkets after the introduction of a farmers market. Sadler, Gilliland, and Arku (2013) further validate this finding, though it was not specific to local foods. As previously mentioned, while the literature almost takes on a "build it they will come" type of mentality, it should also be noted that success is not guaranteed. Kato (2013) stresses this idea in a case study of a failed local foods initiative started in a low-income area within New Orleans. Key takeaways from the Kato study of New Orleans emphasized the importance of producers correctly matching their offerings and prices with local preferences and budgets. Brislen et al. (2015) emphasize the fact that even though a project might be generating numerous streams of community benefit, if the underlying business model will not support the financial sustainability of the enterprise, the project will not survive without external supports.

While there does seem to exist positive evidence that the development of local foods systems helps improve health outcomes for those located in food desert areas, it is fairly narrow in scope. Perhaps most important to note is that there has been no attempt to compare and contrast the costs and benefits of local and nonlocal food policy options and their effectiveness in addressing the issues associated with food deserts. More specifically, why should communities seek local foods vendors instead of a supermarket? While the development of local food systems appears to provide some relief to food deserts, it is not clear if these systems are the most cost effective policy option.

Local foods and social capital

When local foods consumers say that they shop at farmers markets or subscribe to CSAs because they want to "know their farmer," they are talking about building a relationship. In the literature, this relationship is more broadly understood as social capital. The ability of the local food movement to support the creation of social capital is a common argument in favor of supporting local foods (Farley & Bush, 2016). Unfortunately, it is also the least understood. Social capital is a nebulous concept that does not easily lend itself to measurement. It is perhaps for this reason that the local foods and social capital literature has not progressed far beyond the conceptualization, and indeed, the speculation stages. The Schmit, Jablonski, Minner, Kay, and Christensen (2017) article featured elsewhere in this special issue provides an approach and evidence that some forms of intellectual capital (learning) can be transmitted along the networks created by local food systems projects.

To flesh out the relationship between local foods and social capital a bit more, the expectation is that the creation of alternative food networks helps strengthen communities by enhancing trust and fostering cooperation between both consumer and producer participants (Glowacki-Dudka, Murray, & Isaacs, 2013). This building of social capital can occur at the community level and at the farmer level. At the community level, the building of trust would take the form of people (consumers) becoming more aware and networked into the larger community (the food system). At the producer level, social capital is built through farmer networking. Farmer networking has the potential to build viable economic clusters that, ideally, can be self-reinforcing. Value chain concepts would suggest that the exchange of "value" across the supply chain means that social capital is also potentially built between producer and consumer.

Many authors stress that local foods are important for communities because participation in local foods networks can better ground actors in their communities and support social inclusion (Blay-Palmer et al., 2013; DeLind, 2002; Donald & Blay-Palmer, 2006). Further, because local foods are more integrated into the schema of a community and thus better attuned to local deficits, the industry can promote meaningful change and community advancement (Feenstra, 2002; Lyson, 2005; Marsden & Smith, 2005). Local food mechanisms that would have pronounced ability to enhance social capital include community gardens, civic agriculture endeavors, and farm to school programs (Izumi et al., 2010; Lyson, 2005; Teig et al., 2009). In these cases, because actors must work collectively, local foods can be seen as creating a virtuous cycle of social capital building as communities address local concerns (Sonnino & Marsden, 2006). This social capital fostered by local foods systems can then be expended to spill over into other community "sectors."

Surveys support that consumer participation in the local food industry is tied to their civic awareness. Consumers demonstrate interest in the experience of the local food industry, not just the products themselves (Sundbo, 2013). Perhaps more importantly, belief in the societal benefits of their local food purchases mattered more to participation in local foods activity than did cost or ease of access (Knight, 2013). The idea that trust plays an important role in consumer decision-making when it comes to local foods supports the premise that local foods builds social capital (Zagata & Lostak, 2012).

With regard to producer networks, local foods are typified by alternative production schemes that appear to encourage network strengthening. In fact, the desire to build relationships with other producers and consumers is shown to motivate producer participation in Lower Austria (Milestad et al., 2010). Interviews with small farm clusters in the northeastern US anecdotally demonstrate how clusters improve cooperative problem solving and knowledge sharing (Brasier et al., 2007). Results from a survey of wine producers in a region of France also support local foods as promoting cooperative competition (Chiffoleau & Touzard, 2014). It can be said, then, that social capital appears to play a role in producer decision-making.

Unfortunately, despite many arguments supporting the concept and importance of social capital within the local foods context, few authors offer data or empirical analysis to support the claims. Further, while it makes intuitive sense that the development and support of a local foods system can reinforce and perhaps build upon existing social capital, the nature of the interaction remains unclear. Is it that communities that have higher initial levels of social capital are better positioned to create and sustain local food systems, or does the creation of such systems foster the development of social capital? Is there some minimum level of social capital required within the community before a viable local foods system can be developed? Indications on both the consumer and producer sides suggest that a mutual causation might be present, but to what extent, in what ways, and with what impacts, remains largely unexplored.

Critiques of local foods

There are several key limitations characteristic to the arguments advanced by local food movement advocates. As eloquently argued by Born and Purcell (2006), there is a tendency within the local foods literature to presume that if global or large scale is bad, then local or small scale must be good. They refer to this idea as the "local trap." The critiques arise from the weakness of the analytics, skepticism surrounding potential for growth (both of the local foods system itself and its spill over into the larger economy), as well as political concerns. These shortcomings do not necessarily indicate the presence of fallacy in the local foods agenda, rather, they serve as an important and pressing agenda for furthering local foods research and policy analysis. In addition, these shortcomings suggest that communities must move forward with local foods initiatives with open eyes and a cautionary tone.

One key inadequacy of the local foods research is in the analytical support for claims made by local food advocates. One could reasonably argue that there exists a bias throughout the literature in favor of local foods. The result is that much of the literature, particularly the earlier literature, simply assumes truth behind the benefits of the local foods movement rather than supporting claims with data and other empirical evidence (Mount, 2012). For studies that do incorporate data, the bulk of the research relies on case study (Albrecht et al., 2013). In sufficient numbers case studies can provide strong insights, but taken alone, they are generally inconclusive and yield non-generalizable results (Edwards-Jones et al., 2008; Martinez et al., 2010; Tregear, 2011). Because much of the local foods earlier literature is speculative and built on anecdotal accounts, future research should incorporate greater rigor in the testing of suggested ideas (O'Hara & Pirog, 2016; Pearson, Pearson, & Pearson, 2010).

In all fairness, though, it is difficult to rigorously test, defend, and refute ideas if we cannot adequately define what we are studying. This points to the next large issue within the local foods literature, and that is that the very idea of "local" remains unclear. Local remains a relative concept not easily captured or measured (Feagan, 2007). It is defined differently depending on which author is employing the phrase. Further, every definition has a positive connotation attached to the word.

As noted in the introduction, many authors argue against this predisposition of the literature to assume superiority simply because of proximity (Born & Purcell, 2006; Purcell & Brown, 2005; Storey, 2009). The local foods literature essentially equates local foods with fair practice, sustainable production, and tangible economic benefits. With regard to economic benefit, many authors argue that alternative food networks are limited in their ability to grow (Jarosz, 2008). Some authors discuss how promotion of local foods might simply be a renewed expression of "import-substitution," an economic policy often associated with negative consequences (Bellows & Hamm, 2001). To this point, it is becoming clear that producers may not actually achieve higher income levels through means of alternative food network outlets alone (Milestad et al., 2010). Perhaps it would be more productive to focus on some hybrid approach where local foods and larger scale agriculture co-exist and complement one another (Lyson & Guptill, 2004). Given the current level of global integration of economies, it is unreasonable to think that all consumption will switch toward purely local. Local foods must exist within the context of a global economy (Marsden, 2009).

Yet, the literature tends to oversimplifiy the landscape. The rhetoric is overly concerned with discussing successful projects. In fact, in all but two case studies that we are aware of, there are only two incidences of failed initiatives (Brislen et al., 2015; Kato, 2013). The fact that the bulk of the literature tends to overlook failed initiatives is somewhat disappointing because so much can be learned from studying efforts that fell short. Why did the effort fall short? Was it based on unrealistic assumptions? Were there unforeseen implantation problems? Was there a lack of farmer buy-in or were they unable to deliver in a timely and consistent manner or were there simple management mistakes? Insights from failed efforts can actually be more powerful than narrowly focusing on success stories.

With regard to sustainability, it should be noted that local production does not necessarily indicate producers are adhering to sustainable production practices (Winter, 2003). Further, as the previous section highlighted, local production also does not necessarily indicate fair and equitable labor practices (DeLind, 2011; Hinrichs, 2003). This may be a consequence of local politics being susceptible to control by special interest, or it may be due to the fact that clear definition around "fair" and "sustainable" remain elusive (Allen, 2010; DuPuis & Goodman, 2005; Goodman, 2004; Hinrichs & Allen, 2008).

While it is important to remember that the mechanisms employed by the local food movement are still evolving and thus not yet in their final form (Franklin, Newton, & McEntee, 2011), it is equally important to keep in mind the broader claims of the movement and whether or not local foods activity is achieving these goals. There are surely more examples of failed initiatives, initiatives with ambiguous impacts, and local producers who are not beyond reproach. A more balanced portrayal in the literature would help alleviate this problem with bias and ultimately lead to a more intelligent and informed debate as to local foods' best practice. For these reasons, practitioners and planners should be cautious when adopting policy to support the proliferation of local foods before replicable models of production are found to be economically viable.

Conclusion

This review stresses that shortcomings exist in the current literature, particularly the earlier literature, in terms of analytical rigor, in the absence of uniformity in case study analysis, and in the lack of robustness of the results. Local foods systems are seen as a prescription to ailing communities to solve a wide range of societal problems. Yet, rarely does the literature systematically assess impacts. Without the anecdotes and without the rhetoric, the local foods framework is rather flimsy at present. Despite the limitations of much of the earlier literature, the more current work, such as some of the analytical studies in this issue of *Community Development*, appears to be addressing some of these concerns.

Local foods development means developing a food system specific to a community that addresses the needs of producers and consumers. Because the model is tailored to a location, it may not be precisely replicable. Whatever the form of this model, though, the results of such endeavors should have generalizable and robust results. This is particularly true if we are to pursue the promotion of the local foods industry as a strategy of economic development, of community development, and of healthy lifestyles. This is not to say that local food systems do not, in fact, deliver these results in many instances. Rather, up until the past few years we simply have not had sufficiently refined analytical approaches nor the data to provide the clear evidence to support the claims being made by many.

Planners and community development practitioners have a growing role in the creation of efficient and equitable food systems. They need to balance the at times opposing needs of producers and consumers to promote overall community wellbeing. Local foods have quickly become a popular strategy for provisioning for these needs. Yet, widespread adoption of a local foods agenda should be based in decisions made on sound evidence. We should design studies to analyze impacts. We should create industry standards for definitions and measurements. We should allow for critical, unemotional dialogue on the potential impact of local foods, and this dialogue should likely be informed by a host of disciplines and perspectives as suggested by Lamie and colleagues (2014).

As this review suggests, planners and development practitioners should proceed, but they should proceed with caution, especially if investment in policies or programs to support local food system development represents a significant amount of financial, human, or political capital vis a vis other potential investment opportunities. Decision-makers should also consider who is most likely most benefitted by local food system investments. If, over time, the beneficiaries of such investments are from an equity standpoint, deserving, then reasonable arguments can be made that such investments are appropriate, regardless of the additional unclearly defined benefits suggested by advocates. However, given what evidence we currently have of these broader benefits, it would not be wise to rely upon the actual delivery of these benefits as replacements or investments in other areas where these benefits are more likely assured (e.g. health clinics, clean water, and basic access to nutritious food regardless of the source). Decision-makers must strive to optimize their investments with as complete an understanding of the risks and rewards associated with each potential strategy and there is rarely an occasion when these decisions will be made with certainty.

Notes

1. Some care must be taken when looking at the European local food systems literature because of fundamental cultural differences between the Americas and Europe when it comes to food. In the crudest sense, American cuisine could be thought of as a drive-thru fast food restaurant, while in Europe each meal is an event to be savored. Though, one could argue that in many parts of Western Europe the introduction of fast food establishments is changing the relationship Europeans have with food.
2. For more insights and detailed data on food hubs see http://www.ngfn.org/resources/food-hubs/benchmarking-study/press-release-food-hub-benchmarking-study.
3. USDA Food Atlas: https://www.ers.usda.gov/data-products/food-environment-atlas/data-access-and-documentation-downloads/.
4. See New York Times article at http://www.nytimes.com/2013/05/26/opinion/sunday/breeding-the-nutrition-out-of-our-food.html?pagewanted=all&_r=0.
5. For a more detailed set of materials and case-studies see https://localfoodeconomics.com/.

Disclosure statement

No potential conflict of interest was reported by the authors.

Funding

This work was supported in part by the Wisconsin Agricultural Experiment Station, University of Wisconsin-Madison and the Southern Rural Development Center, Mississippi State University. Technical Contribution No. 6581 of the Clemson University Experiment Station, College of Agriculture, Forestry and Life Sciences.

References

Adalja, A., Hanson, J., Towe, C., & Tselepidakis, E. (2015). An examination of consumer willingness to pay for local products. *Agricultural and Resource Economics Review, 44*, 253–274. doi:10.1017/S1068280500005050

Adams, D.C., & Salois, M.J. (2010). Local versus organic: A turn in consumer preferences and willingness-to-pay. *Renewable Agriculture and Food Systems, 25*, 331–341. doi:10.1017/S1742170510000219

Ahearn, M., & Sterns, J. (2013). Direct-to-consumer sales of farm products: Producers and supply chains in the Southeast. *Journal of Agricultural and Applied Economics, 45*, 497–508.

Albrecht, C., Johnson, R., Hamann, S., Sneyd, L., Ohberg, L., & CoDye, M. (2013). Toward alternative food systems development: Exploring limitations and research opportunities. *Journal of Agriculture, Food Systems, and Community Development, 3*, 151–159. doi:10.5304/jafscd.2013.034.019

Allen, P. (2010). Realizing justice in local food systems. *Cambridge Journal of Regions, Economy and Society, 3*, 295–308. doi:10.1093/cjres/rsq015

Armstrong, D. (2000). A survey of community gardens in upstate New York: Implications for health promotion and community development. *Health & Place, 6*, 319–327. doi:10.1016/S1353-8292(00),00013-7

Bader, M.D.M., Purciel, M., Yousefzaseh, P., & Neckerman, K.M. (2010). Disparities in neighborhood food environments: Implications of measurement strategies. *Economic Geography, 86*, 409–430. doi:10.1111/j.1944-8287.2010.01084.x

Ballingall, J., & Winchester, N. (2010). Food miles: Starving the poor? *The World Economy, 33*, 1201–1217. doi:10.1111/j.1467-9701.2010.01270.x

Barham, E., Bingen, J., Lovenworth, S., Kole, P., & Mendelson, R. (2010). *American origin products: Protecting a legacy*. Geneva: oriGin.

Barham, J., Tropp, D., Enterline, K., Farbman, J., Fisk, J., & Kiraly, S. (2012). *Regional food hub resource guide*. Washington, DC: US Department of Agriculture, Agricultural Marketing Service. doi:10.9752/MS046.04-2012

Barnett, C., Clarke, N., Cloke, P., & Malpass, A. (2005). The political ethics of consumerism. *Consumer Policy Review, 15*, 45–51. Retrieved from http://oro.open.ac.uk/id/eprint/7143

Bellows, A.C., & Hamm, M.W. (2001). Local autonomy and sustainable development: Testing import substitution in more localized food systems. *Agriculture and Human Values, 18*, 271–284. doi:10.1023/A:1011967021585

Berlin, L., Norris, K., Kolodinsky, J., & Nelson, A. (2013). The role of social cognitive theory in farm-to-school-related activities: Implications for child nutrition. *Journal of School Health, 83*, 589–595. doi:10.1111/josh.12069

Berning, J.P. (2012). Access to local agriculture and weight outcomes. *Agricultural and Resource Economics Review, 41*, 57–71. doi:10.1017/S1068280500004184

Bimbo, F., R. Viscecchia, & G. Nardone. (2012, June). Does the alternative food supply network affect the human health. In *126th Seminar, June 27–29, 2012, Capri, Italy* (No. 126060). European Association of Agricultural Economists.

Blay-Palmer, A., Landman, K., Knezevic, I., & Hayhurst, R. (2013). Constructing resilient, transformative communities through sustainable "food hubs". *Local Environment, 18*, 521–528. doi:10.1080/13549839.2013.797156

Bloom, J.D., & Hinrichs, C.C. (2011). Moving local food through conventional food system infrastructure: Value chain framework comparisons and insights. *Renewable Agriculture and Food Systems, 26*, 13–23. doi:10.1017/S1742170510000384

Born, B., & Purcell, M. (2006). Avoiding the local trap scale and food systems in planning research. *Journal of Planning Education and Research, 26*, 195–207. doi:10.1177/0739456X06291389

Bowen, S., & Mutersbaugh. (2014). Local or localized? Exploring the contributions of Franco-Mediterranean agrifood theory to alternative food research. *Agriculture and Human Values, 31*, 201–213.

Boys, K.A., & Hughes, D.W. (2013). A regional economics–based research agenda for local food systems. *Journal of Agriculture, Food Systems, and Community Development, 3*, 145–150. doi:10.5304/jafscd.2013.034.012

Brasier, K.J., Goetz, S., Smith, L.A., Ames, M., Green, J., Kelsey, T., … Whitmer, W. (2007). Small farm clusters and pathways to rural community sustainability. *Community Development, 38*, 8–22. doi:10.1080/15575330709489826

Brislen, L., Woods, T., Meyer, L., & Routt, N. (2015). *Grasshoppers distribution: Lessons learned and a lasting legacy*. SR-108. Lexington, KY: University of Kentucky Agricultural Experiment Station.

Brown, J.P., Goetz, S.J., Ahearn, M.C., & Laing, C.K. (2014). Linkages between community-focused agriculture, farm sales, and regional Growth. *Economic Development Quarterly, 28*, 5–16. doi:10.1177/0891242413506610

Brown, C., & Miller, S. (2008). The impacts of local markets: A review of research on farmers' markets and community supported agriculture (CSA). *American Journal of Agricultural Economics, 90*, 1298–1302. doi:10.1111/j.1467-8276.2008.01220.x

Byker, C.J., Misyak, S., Shanks, J., & Serrano, E.L. (2013). Do farmers' markets improve diet of participants using federal nutrition assistance programs? A literature review. *Journal of Extension, 51*(6). Retrieved from https://www.joe.org/joe/2013december/a5.php

Campbell, D.C., Carlisle-Cummins, I., & Feenstra, G. (2013). Community food systems: Strengthening the research to practice continuum. *Journal of Agriculture, Food Systems, and Community Development, 3*, 121–138. Retrieved from https://www.foodsystemsjournal.org/index.php/fsj/article/view/171

Carpio, C.E., & Isengildina-Massa, O. (2009). Consumer willingness to pay for locally grown products: The case of South Carolina. *Agribusiness, 25*, 412–426. doi:10.1002/agr.20210

Carson, R.A., Hamel, Z., Giarrocco, K., Baylor, R., & Mathews, L.G. (2016). Buying in: The influence of interactions at farmers' markets. *Agriculture and Human Values, 33*, 861–875. Retrieved from https://link.springer.com/article/10.1007/s10460-015-9675-y

Chamberlain, A.J., Kelley, K.M., & Hyde, J. (2013). Mid-Atlantic consumer purchasing behavior and knowledge of locally grown and seasonal produce. *Journal of Extension, 51*(2). Retrieved from https://www.joe.org/joe/2013april/rb4.php

Chiffoleau, Y., & Touzard, J.M. (2014). Understanding local agri-food systems through advice network analysis. *Agriculture and Human Values, 31*, 19–32. Retrieved from https://link.springer.com/article/10.1007/s10460-013-9446-6

Christensen, B., & Phillips, R. (2016). Local food systems and community economic development through the lens of theory. *Community Development, 47*, 638–651. doi:10.1080/15575330.2016.1214609

Christy, E., Landman, K., Nowatschin, E., & Blay-Palmer, A. (2013). *Local food systems in North America*. Retrieved from http://nourishingontario.ca/wp-content/uploads/2013/07/NA-FoodHub-LitReview-2013.pdf

Cleveland, D.A., Muller, N.M., Tranovick, A.C., Mazaroli, D.N., & Hinson, K. (2014). Local food hubs for alternative food systems: A case study from Santa Barbara County, California. *Journal of Rural Studies, 35*, 26–36. doi:10.1016/j.jrurstud.2014.03.008

Cranfield, J., Henson, S., & Blandon, J. (2012). The effect of attitudinal and sociodemographic factors on the likelihood of buying locally produced food. *Agribusiness, 28*, 205–221. doi:10.1002/agr.21291

DeLind, L.B. (2002). Place, work, and civic agriculture: Common fields for cultivation. *Agriculture and Human Values, 19*, 217–224. doi:10.1023/A:1019994728252

DeLind, L.B. (2011). Are local food and the local food movement taking us where we want to go? Or are we hitching our wagons to the wrong stars? *Agriculture and Human Values, 28*, 273–283. doi:10.1007/s10460-010-9263-0

Deller, S.C., Brown, L., & Canto, A. (2014). *Local foods and community health: An exploratory analysis.* Department of Agricultural and Applied Economics, University of Wisconsin-Madison Staff Paper No. 571. Retrieved from http://www.aae.wisc.edu/pubs/sps/pdf/stpap571.pdf

Deller, S.C., Brown, L., Haines, A., & Fortenbery, R. (2014). *Local foods and rural economic growth.* Department of Agricultural and Applied Economics, University of Wisconsin-Madison Staff Paper No. 570. Retrieved from http://www.aae.wisc.edu/pubs/sps/pdf/stpap570.pdf

Deller, S. C., Canto, A., & Brown, L. (2017). Food access, local foods, and community health. *Community Development,* 1–24.

DeWeerdt, S. (2009). Local food: The economics. *World Watch, 22*, 20–24. Retrieved from http://www-personal.umich.edu/~prestos/Downloads/DC/Schoolman_DeWeerdt2009b.pdf

Diamond, A., Tropp, D., Barham, J., Muldoon, M.F., Kiraly, S., & Cantrell, P. (2014). *Food value chains: Creating shared value to enhance marketing success.* US Department of Agriculture, Agricultural Marketing Service. Retrieved from http://ageconsearch.umn.edu/record/176263

Dolstad, H.A., Woodward, A.R., Green, C.L., & Mcspirit, S.J. (2016). Interest in nutrition and local food systems among food-insecure households in an Appalachian community. *Journal of Hunger & Environmental Nutrition, 11*, 340–353. doi:10.1080/19320248.2016.1146191

Donald, B. (2013). Food retail and access after the crash: Rethinking the food desert problem. *Journal of Economic Geography, 13*, 231–237. doi:10.1093/jeg/lbs064

Donald, B., & Blay-Palmer, A. (2006). The urban creative-food economy: Producing food for the urban elite or social inclusion opportunity? *Environment and Planning A, 38*, 1901–1920. doi:10.1068/a37262

DuPuis, E.M., & Goodman, D. (2005). Should we go "home" to eat? Toward a reflexive politics of localism. *Journal of Rural Studies, 21*, 359–371. doi:10.1016/j.jrurstud.2005.05.011

Eaton, E. (2008). From feeding the locals to selling the locale: Adapting local sustainable food projects in Niagara to neocommunitarianism and neoliberalism. *Geoforum, 39*, 994–1006. doi:10.1016/j.geoforum.2007.10.017

Edwards-Jones, G., Canals, L.M., Hounsome, N., Truninger, M., Koerber, G., Hounsome, B., … Jones, D.L. (2008). Testing the assertion that 'local food is best': The challenges of an evidence-based approach. *Trends in Food Science & Technology, 19*, 265–274. doi:10.1016/j.tifs.2008.01.008

Farley, K.W., & Bush, C.B. (2016). Using relationships as resources in social impact investing: Examining a local food movement in Appalachia. *Journal of Appalachian Studies, 22*, 224–244. doi:10.5406/jappastud.22.2.0224

Feagan, R. (2007). The place of food: Mapping out the 'local' in local food systems. *Progress in Human Geography, 31*, 23–42. doi:10.1177/0309132507073527

Feenstra, G. (2002). Creating space for sustainable food systems: Lessons from the field. *Agriculture and Human Values, 19*, 99–106. doi:10.1023/A:1016095421310

Franklin, A., Newton, J., & McEntee, J.C. (2011). Moving beyond the alternative: Sustainable communities, rural resilience and the mainstreaming of local food. *Local Environment, 16*, 771–788. doi:10.1080/13549839.2011.574685

Garnett, T. (2013). Food sustainability: Problems, perspectives and solutions. *Proceedings of the Nutrition Society, 72*, 29–39. doi:10.1017/S0029665112002947

Glowacki-Dudka, M., Murray, J., & Isaacs, K.P. (2013). Examining social capital within a local food system. *Community Development Journal, 48*, 75–88. doi:10.1093/cdj/bss007

Goodman, D. (2004). Rural Europe redux? Reflections on alternative agro-food networks and paradigm change. *Sociologia Ruralis, 44*, 3–16. doi:10.1111/j.1467-9523.2004.00258.x

Goodman, M.K. (2013). Wrapping and stuffing food relationally: Pleasure, place, production and power. In M. McWilliams (Ed.), *Oxford symposium on food and cookery: Wrapped and stuffed foods* (pp. 52–56). Devon: Prospect Books.

Green, J.J., Worstell, J., & Canarios, C. (2017). Local Agrifood System Sustainability/Resilience Index (SRI): Constructing a data tool applied to counties in the southern United States. *Community Development, 48*.

Grewal, S.S., & Grewal, P.S. (2012). Can cities become self-reliant in food? *Cities, 29*, 1–11. doi:10.1016/j.cities.2011.06.003

Hendrickson, M. (2009). Creating alternatives: A participant observer's reflections on the emerging local food system in Kansas City. *Southern Rural Sociology, 24*, 169–191. Retrieved from http://journalofruralsocialsciences.org/pages/Articles/SRS%202009%2024/2/SRS%202009%2024%202%20169-191.pdf

Henneberry, S.R., Whitacre, B., & Agustini, H.N. (2009). An evaluation of the economic impacts of Oklahoma farmers' markets. *Journal of Food Distribution Research, 40*, 64–78. Retrieved from http://ageconsearch.tind.io/record/99760/files/Evaluation%20 pg%2064-78.pdf

Hightower, L.S., & Brennan, M.A. (2013). *Local food systems, ethnic entrepreneurs, and social networks.* Agricultural and Applied Economics Association. Retrieved from http://ageconsearch.tind.io/record/149696/files/Local%20Food%20Systems_%20Ethnic%20Entrepreneurs_%20and%20Social%20Networks%20-%20Hightower_L.pdf

Hinrichs, C.C. (2000). Embeddedness and local food systems: Notes on two types of direct agricultural market. *Journal of Rural Studies, 16*, 295–303. doi:10.1016/S0743-0167(99),00063-7

Hinrichs, C.C. (2003). The practice and politics of food system localization. *Journal of Rural Studies, 19*, 33–45. doi:10.1016/S0743-0167(02),00040-2

Hinrichs, C.C., & Allen, P. (2008). Selective patronage and social justice: Local food consumer campaigns in historical context. *Journal of Agricultural and Environmental Ethics, 21*, 329–352. doi:10.1007/s10806-008-9089-6

Holben, D.H. (2010). Farmers markets: Fertile ground for optimizing health. *Journal of the American Dietetic Association, 110*, 364–365. doi:10.1016/j.jada.2009.11.015

Holt, G., & Amilien, V. (2007). Introduction: From local food to localised food. *Anthropology of Food*, S2. Retrieved from https://aof.revues.org/405

Hughes, D.W., Brown, C., Miller, S., & McConnell, T. (2008). Evaluating the economic impact of farmers' markets using an opportunity cost framework. *Journal of Agricultural and Applied Economics, 40*, 253–265. doi:10.1017/S1074070800028091

Izumi, B.T., Wright, D.W., & Hamm, M.W. (2010). Market diversification and social benefits: Motivations of farmers participating in farm to school programs. *Journal of Rural Studies, 26*, 374–382. doi:10.1016/j.jrurstud.2010.02.002

Jablonski, B.B.R., Schmit, T.M., & Kay, D. (2016). Assessing the economic impacts of food hubs on regional economies: A framework that includes opportunity cost. *Agricultural and Resource Economics Review, 45*, 143–172. doi:10.1017/age.2016.9

Jarosz, L. (2008). The city in the country: Growing alternative food networks in metropolitan areas. *Journal of Rural Studies, 24*, 231–244. doi:10.1016/j.jrurstud.2007.10.002

Kato, Y. (2013). Not just the price of food: Challenges of an urban agriculture organization in engaging local residents. *Sociological Inquiry, 83*, 369–391. doi:10.1111/soin.12008

Knezevic, I., Landman, K., & Blay-Palmer, A. (2013). *Local food systems–International perspectives: A review.* Ontario Ministry of Agriculture, Food and Rural Affairs, Canada. Retrieved from http://www.glamur.eu/wp-content/uploads/2013/07/EUAntipode-FoodHub-LitReview-2013.pdf

Knight, A.J. (2013). Evaluating local food programs: The case of select Nova Scotia. *Evaluation and Program Planning, 36*, 29–39. doi:10.1016/j.evalprogplan.2012.05.003

Kremer, P., & DeLiberty, T.L. (2011). Local food practices and growing potential: Mapping the case of Philadelphia. *Applied Geography, 31*, 1252–1261. doi:10.1016/j.apgeog.2011.01.007

Lamie, R.D., Dunning, R., Bendfeldt, E., Lelekacs, J.M., Velandia, M., & Meyer, L. (2014). Local food systems in the South: A call for a collaborative approach to assessment. *Choices, 28*(4). Retrieved from http://www.choicesmagazine.org/UserFiles/file/cmstheme_342.pdf

Larsen, K., & Gilliland, J. (2009). A farmers' market in a food desert: Evaluating impacts on the price and availability of healthy food. *Health & Place, 15*, 1158–1162. doi:10.1016/j.healthplace.2009.06.007

Little, R., Maye, D., & Ilbery, B. (2010). Collective purchase: Moving local and organic foods beyond the niche market. *Environment and Planning A, 42*, 1797–1813. doi:10.1068/a4262

Low, S.A., Adalja, A., Beaulieu, E., Key, N., Martinez, S., Melton, A., … Jablonski, B.B.R. (2015). *Trends in US local and regional food systems.* United States Department of Agriculture, Economic Research Service, Administrative Publication 068. Retrieved from https://naldc.nal.usda.gov/download/60312/PDF

Low, S.A., & Vogel, S.J. (2011). *Direct and intermediated marketing of local foods in the United States.* United States Department of Agriculture, Economic Research Service. Retrieved from https://www.ers.usda.gov/webdocs/publications/44924/8276_err128_2_.pdf?v=41056

Lyson, T.A. (2005). Civic agriculture and community problem solving. *Culture & Agriculture, 27*, 92–98. doi:10.1525/cag.2005.27.2.92

Lyson, T.A., & Guptill, A. (2004). Commodity agriculture, civic agriculture and the future of US farming. *Rural Sociology, 69*, 370–385. doi:10.1526/0036011041730464

Maples, M., Morgan, K.L., Interis, M.G., & Harri, A. (2013). Who buys food directly from producers in the southeastern United States? *Journal of Agricultural and Applied Economics, 45*, 509–518. Retrieved from https://ageconsearch.umn.edu/bitstream/142929/2/SAEA2013_Maplesetal_JAAE_corrected.pdf10.1017/S1074070800005022

Marsden, T. (2009). Mobilities, vulnerabilities and sustainabilities: Exploring pathways from denial to sustainable rural development. *Sociologia Ruralis, 49*, 113–131. doi:10.1111/j.1467-9523.2009.00479.x

Marsden, T., & Smith, E. (2005). Ecological entrepreneurship: Sustainable development in local communities through quality food production and local branding. *Geoforum, 36*, 440–451. doi:10.1016/j.geoforum.2004.07.008

Martinez, S., Hand, M.S., Da Pra, M., Pollack, S., Ralston, K., Smith, T., … Newman, C. (2010). *Local food systems: Concepts, impacts, and issues*. Economic Research Report 97. US Department of Agriculture, Economic Research Service. Retrieved from https://www.ers.usda.gov/webdocs/publications/46393/7054_err97_1_.pdf?v=42265

Matson, J., Sullins, M., & Cook, C. (2013). *The role of food hubs in local food marketing*. USDA Rural Development Service Report 73. United States Department of Agriculture, Economic Research Service. Retrieved from https://www.rd.usda.gov/files/sr73.pdf

Matson, J., & Thayer, J. (2016). The role of food hubs in food supply chains. *Journal of Agriculture, Food Systems, and Community Development, 3*, 43–47. doi:10.5304/jafscd.2013.034.004

McCaffrey, S. J., & Kurland, N. (2014). Who defines "local"? Resistance to harmonizing standards in ethical markets. *Business and Politics, 16*, 191–219.

McFadden, D.T. (2015). What do we mean by local foods? *Choices, 30*, 1–6. Retrieved from http://www.jstor.org/stable/choices.30.1.02

McFadden, D.T., Bond, C.A., & Bond, J.K. (2008). Going local: Exploring consumer behavior and motivations for direct food purchases. *American Journal of Agricultural Economics, 90*, 1303–1309. doi:10.1111/j.1467-8276.2008.01221.x

McFadden, D.T., Conner, D., Deller, S., Hughes, D., Meter, K., Morales, A., & Tropp, D. (2016). *The economics of local food systems: A toolkit to guide community discussions, assessments and choices.* US Department of Agriculture, Agricultural Marketing Service. Retrieved from https://www.rd.usda.gov/files/ILAMSToolkit.pdf

Milestad, R., Bartel-Kratochvil, R., Leitner, H., & Axmann, P. (2010). Being close: The quality of social relationships in a local organic cereal and bread network in Lower Austria. *Journal of Rural Studies, 26*, 228–240. doi:10.1016/j.jrurstud.2010.01.004

Mount, P. (2012). Growing local food: Scale and local food systems governance. *Agriculture and Human Values, 29*, 107–121.

National Grocers Association. (2015). National Grocers Association Supermarketguru® consumer survey report. Retrieved from https://tinyurl.com/gnwgl56

Nguyen, T.-V., Wysocki, A., Treadwell, D., Farnsworth, D., & Clark, J. (2008). *Economics of the organic food industry in Florida*. Gainesville: University of Florida, IFAS Extension, publication no. FE732. Retrieved from http://edis.ifas.ufl.edu/fe732

Oberholtzer, L., Hanson, J.C., Brust, G., Dimitri, C., & Richman, N. (2012). Local foods in Maryland schools and implications for extension: Findings from schools and farmers. *Journal of Extension, 50*(2), 1–9. Retrieved from https://www.joe.org/joe/2012april/rb4.php

O'Hara, S. (2017). The urban food hubs solution: Building capacity in urban communities. *Metropolitan Universities, 28*, 69–93 doi:10.18060/21477

O'Hara, J.K., & Pirog, R. (2016). Economic impacts of local food systems: Future research priorities. *Journal of Agriculture, Food Systems, and Community Development, 3*, 35–42.

Pearson, L.J., Pearson, L., & Pearson, C.J. (2010). Sustainable urban agriculture: Stocktake and opportunities. *International Journal of Agricultural Sustainability, 8*, 7–19. doi:10.3763/ijas.2009.0468

Pelletier, J.E., Laska, M.N., Neumark-Sztainer, D., & Story, M. (2013). Positive attitudes toward organic, local, and sustainable foods are associated with higher dietary quality among young adults. *Journal of the Academy of Nutrition and Dietetics, 113*, 127–132. doi:10.1016/j.jand.2012.08.021

Peterson, H.H., Barkley, A., Chacón-Cascante, A., & Kastens, T.L. (2012). The motivation for organic grain farming in the United States: Profits, lifestyle, or the environment? *Journal of Agricultural and Applied Economics, 44*, 137–155.

Pinchot, A. (2014). *The economics of local food systems: A literature review of the production, distribution, and consumption of local food*. University of Minnesota-Extension, Extension Center for Community Vitality. Retrieved from http://conservancy.umn.edu/bitstream/handle/11299/171637/2014-Economics-of-Local-Food-Systems.pdf?sequence=1&isAllowed=y

Purcell, M., & Brown, J.C. (2005). Against the local trap: Scale and the study of environment and development. *Progress in Development Studies, 5*, 279–297. doi:10.1191/1464993405ps122oa

Ralston, K., & Benson, M. (2015). *Many US school districts serve local foods*. US Department of Agriculture, Economic Research Service. Retrieved from https://www.ers.usda.gov/amber-waves/2015/march/many-us-school-districts-serve-local-foods/

Renting, H., Marsden, T.K., & Banks, J. (2003). Understanding alternative food networks: Exploring the role of short food supply chains in rural development. *Environment and Planning A, 35*, 393–411. doi:10.1068/a3510

Ruge, D., & Mikkelsen, B.E. (2013). Local public food strategies as a social innovation: Early insights from the LOMA-Nymarkskolen case study. *Acta Agriculturae Scandinavica, Section B – Soil & Plant Science, 63*, 56–65.

Sadler, R.C., Clark, M.A.R., & Gilliland, J.A. (2013). An economic impact comparative analysis of farmers markets in Michigan and Ontario. *Journal of Agriculture, Food Systems, and Community Development, 3*, 61–81. doi:10.5304/jafscd.2013.033.009

Sadler, R.C., Gilliland, J.A., & Arku, G. (2013). Community development and the influence of new food retail sources on the price and availability of nutritious food. *Journal of Urban Affairs, 35*, 471–491. doi:10.1111/j.1467-9906.2012.00624.x

Salois, M.J. (2012). Obesity and diabetes, the built environment, and the 'local' food economy in the United States, 2007. *Economics & Human Biology, 10*, 35–42. doi:10.1016/j.ehb.2011.04.001

Schmit, T.M., Jablonski, B.B.R., Minner, J., Kay, D., & Christensen, L. (2017). Rural wealth creation of intellectual capital from urban local food system initiatives: Developing indicators to assess change. *Community Development, 48*.

Sharp, J.S., Clark, J.K., Davis, G.A., Smith, M.B., & McCutcheon, J.S. (2011). Adapting community and economic development tools to the study of local foods: The case of Knox County, Ohio. *Journal of Extension, 49*(2), 1–12. Retrieved from https://www.joe.org/joe/2011april/pdf/JOE_v49_2a4.pdf

Sonnino, R., & Marsden, T. (2006). Beyond the divide: Rethinking relationships between alternative and conventional food networks in Europe. *Journal of Economic Geography, 6*, 181–199. doi:10.1093/jeg/lbi006

Stagl, S. (2002). Local organic food markets: Potentials and limitations for contributing to sustainable development. *Empirica, 29*, 145–162. doi:10.1023/A:1015656400998

Storey, D. (2009). Rural economic development. *International Encyclopedia of Human Geography*, 310–314. doi:10.1016/B978-008044910-4.00898-1

Sundbo, D.I.C. (2013). Local food: The social construction of a concept. *Acta Agriculturae Scandinavica, Section B – Soil & Plant Science, 63*, 66–77.

Swenson, D.A. (2011, March 25). *The regional economic development potential and constraints to local foods development in the Midwest*. Presented at the 50th Annual Meeting of the Southern Regional Science Association, New Orleans.

Taylor, D.F., & Miller, C.R. (2010). Rethinking local business clusters: The case of food clusters for promoting community development. *Community Development, 41*, 108–120. doi:10.1080/15575330903548778

Teig, E., Amulya, J., Bardwell, L., Buchenau, M., Marshall, J.A., & Litt, J.S. (2009). Collective efficacy in Denver, Colorado: Strengthening neighborhoods and health through community gardens. *Health & Place, 15*, 1115–1122. doi:10.1016/j.healthplace.2009.06.003

Telligman, A.L., Worosz, M.R., & Bratcher, C.L. (2017). "Local" as an indicator of beef quality: An exploratory study of rural consumers in the southern US. *Food Quality and Preference, 57*, 41–53. doi:10.1016/j.foodqual.2016.11.001

Thilmany, D., Conner, D., Curtis, K.R., Liang, C.L.K., Mulik, K., O'Hara, J., & Woods, T. (2013). Researching market and supply-chain opportunities for local foods systems: Setting priorities and identifying linkages. *Journal of Agriculture, Food Systems, and Community Development, 3*, 131–137. doi:10.5304/jafscd.2013.034.018

Tregear, A. (2011). Progressing knowledge in alternative and local food networks: Critical reflections and a research agenda. *Journal of Rural Studies, 27*, 419–430. doi:10.1016/j.jrurstud.2011.06.003

Walker, R.E., Keane, C.R., & Burke, J.G. (2010). Disparities and access to healthy food in the United States: A review of food deserts literature. *Health & Place, 16*, 876–884. doi:10.1016/j.healthplace.2010.04.013

Webber, K.H., Stephenson, T.J., Mayes, L., & Stephenson, L. (2013). Nutrition knowledge and dietary habits of farmers markets patrons. *World Applied Sciences Journal, 23*, 267–271.

Willis, D.B., Carpio, C.E., Boys, K.A., & Young, E.D. (2013, August 4–6). *Consumer willingness to pay for locally grown produce designed to support local food banks and enhance locally grown producer markets*. Agricultural and Applied Economics Association, 2013. Selected paper prepared for presentation at the Agricultural & Applied Economics Association's 2013 AAEA & CAES Joint Annual Meeting, Washington, DC.

Winter, M. (2003). Embeddedness, the new food economy and defensive localism. *Journal of Rural Studies, 19*, 23–32. doi:10.1016/S0743-0167(02)00053-0

Wirth, F.F., Stanton, J.L., & Wiley, J.B. (2011). The relative importance of search versus credence product attributes: Organic and locally grown. *Agricultural and Resource Economics Review, 40*, 48–62.

Wittman, H., Beckie, M., & Hergesheimer, C. (2012). Linking local food systems and the social economy? Future roles for farmers markets in Alberta and British Columbia. *Rural Sociology, 77*, 36–61. doi:10.1111/j.1549-0831.2011.00068.x

Woods, M. (2012). Rural geography III: Rural futures and the future of rural geography. *Progress in Human Geography, 36*, 125–134. doi:10.1177/0309132510393135

Woods, T., Ernst, M., & Tropp, D. (2017). *Community supported agriculture: New models for changing markets*. Washington, DC: US Department of Agriculture, Agricultural Marketing Service.

Woods, T., Rossib, J., & Davis, A. (2017). Local food vitality index: Measuring consumer attitudes toward food system attributes. *Journal of Food Distribution Research, 48*, 33–41.

Zagata, L., & Lostak, M. (2012). In goodness we trust. The role of trust and institutions underpinning trust in the organic food market. *Sociologia Ruralis, 52*, 470–487. doi:10.1111/j.1467-9523.2012.00574.x

Zepeda, L., & Nie, C. (2012). What are the odds of being an organic or local food shopper? Multivariate analysis of US food shopper lifestyle segments. *Agriculture and Human Values, 29*, 467–480. doi:10.1007/s10460-012-9364-z

Rural wealth creation of intellectual capital from urban local food system initiatives: Developing indicators to assess change

Todd M. Schmit, Becca B. R. Jablonski, Jennifer Minner, David Kay, and Libby Christensen

ABSTRACT

With growing interest in wealth creation in rural areas from farmer participation in urban local food markets, defining empirical measures is crucial. However, data limitations and a lack of agreement on what serves as reasonable proxies for alternative capital measures makes the process challenging. Using the Delphi Method and considering a large urban farmers market system, this study prioritizes impacts on eight forms of community capital and estimates empirically flows of intellectual capital to rural areas using primary data from market vendors and customers. Results indicate that sufficient engagement between farmers and urban consumers, combined with educational programming by the farmers market, can increase the transmission of intellectual capital flows to rural areas. For the large urban-based farmers market evaluated, we find that such participation is associated with higher human capital stocks in the counties in which the farmers reside.

In a number of dimensions, rural people and places in the United States have fared worse than their urban counterparts. Traditionally, rural areas have experienced higher poverty rates and lower educational attainment. Since 2010, rural areas have been witnessing declining populations as urban populations continue to grow (Kusmin, 2015). At the community level, this can result in lower demands for jobs, decreases in workforce quality, and increases in the per capita costs of providing public services (Jablonski, 2014a).

Under the Obama Administration, the US Department of Agriculture (USDA) promoted local and regional food systems development as a key strategy in supporting agricultural and rural economic development (USDA, 2015). Efforts through private foundations and development organizations echoed this theme. Between 2009 and 2014, USDA invested over $1 billion in local and regional food systems projects (Vilsack, 2016). Though these initiatives are likely to shift moving forward, the Trump administration is looking for opportunities to support rural communities through agricultural initiatives (Trump, 2017). Given research demonstrating growing consumer demand for differentiated food products, there also exists new opportunities to link food systems development with rural community development (Thilmany, Bond, & Bond, 2008).

Understanding the rural impacts of local food system investments is relatively limited in focus, generally considering impacts on employment, economic output, or labor income (Gunter & Thilmany, 2012; Hughes, Brown, Miller, & McConnell, 2008; Hughes & Isengildina-Massa, 2015; Jablonski, Schmit, & Kay, 2016; Schmit, Jablonski, & Mansury, 2016; Swenson, 2010). A toolkit assisting communities in conducting these types of assessments has recently been released by USDA (Thilmany McFadden et al., 2016). Other approaches examining impacts of local food system activity utilized spatial econometric models, where a measure of local foods activity like direct-to-consumer sales is used as an explanatory variable in describing changes in economic growth such as per capita income (Brown, Goetz, Ahearn, & Liang, 2014; Deller, Brown, Haines, & Fortenbery, 2014). Collectively, these studies generally show relatively small short-term gains accruing to local economies.

A more comprehensive evaluation of community impacts from local food system activity is necessary to better inform the efficacy of development efforts and guide policy decisions. Lamie et al. (2013) promote collaborative approaches to local food systems assessments that incorporate a range of community stakeholders and outcomes. Recent research promoting the concept of rural wealth creation (RWC) considers a broader set of impacts concerning community wealth (Pender, Marre, & Reeder, 2012a, 2012b), but little has been done in the context of local food systems. Only Jablonski (2014b) provides a discussion of RWC as applied to urban farmers markets through a comparative literature review, but without any empirical measurement.

In the RWC context, wealth includes all types of community capital assets that contribute to the wellbeing of people and communities, often categorized into social, cultural, individual, intellectual, political, physical, natural, and financial capitals (Pender, Johnson, Weber, & Fannin, 2014). The concept of broader community wealth measures is not new, of course; however, the application of these concepts to economic development has gained renewed attention in and beyond academia, including, for example, the World Bank (2006, 2011) and the United Nations (United National Environmental Programme & Secretariat of the International Human Dimensions Programme on Global Environmental Change [UNU-IHDP & UNEP], 2012).

RWC draws heavily from the Community Capitals Framework (CCF) developed by Flora and Flora (2008). The CCF provides for enhanced understanding of the processes underlying community development through describing capitals that can be used individually or in combination to produce community change (Emery & Flora, 2006; Flora & Flora, 2008; Green & Haines, 2016). The concept of RWC is, in this sense, complementary to the CCF, and further promotes efforts to develop empirically tractable indicators of community capital with rigorous testing of the relationship of these indicators to actual community outcomes (Pender et al., 2012b, 2014).[1]

Evaluating community outcomes relative to the levels of (or changes in) capital have generally focused on a specific capital, e.g. Goetz and Rupasingha (2006), Rupasingha, Goetz, and Freshwater (2006), and Putnam (2007) with social capital, and Arrow, Dasgupta, Goulder, Mumford, and Oleson (2012) and Fenichel and Abbott (2014) with natural capital. Emery and Flora (2006) argue that while communities may differ in stocks of capital, key capitals, such as human, social, and financial, are likely to positively influence others. That said, little empirical evidence exists that describes how the investment in one capital may be related to and/or affect the others (Pigg, Gasteyer, Martin, Keating, & Apaliyah, 2013). This issue is highlighted in Johnson, Raines, and Pender (2014), where they develop an extended social accounting matrix (SAM) model that incorporates community capital assets and the interactions between them and traditional sector accounts. However, they note that data and measurement issues are challenging, particularly in populating the large number of parameters within a SAM associated with the capital relationships. With an alternative methodological orientation, Brown and Lewin (2016) and Pender (2015) propose proxy variables for multiple capital measures, on a relatively ad hoc basis, to identify relationships with per capita incomes and land values, respectively. The challenge remains in how best to prioritize capital impacts and associate them with empirically tractable indicators.

To address these concerns, this study provides three important contributions. First, we recommend a set of prioritized impacts for eight capital assets in the context of RWC and urban farmers markets using the Delphi Method (DM). Second, we discuss in detail the associated indicators identified from the DM for intellectual capital, i.e. the stock of knowledge, innovation, and creativity or imagination in a region. While this restricts our ability to assess interactive capital effects, we do this for three primary reasons: (1) impacts on intellectual capital were featured prominently in the advisory panel discussions related to the size of urban farmers market initiative considered and the educational efforts they conduct, (2) human capital (of which intellectual capital is a component) is considered an ideal entry point for community development and key in influencing the levels of other capital (Emery & Flora, 2006), and (3) an index of the stock of human capital we derive from secondary data adds to the richness of the discussion when compared with the primary data analysis. Our third contribution is empirical. Using the indicators identified by the DM, we describe intellectual capital flows to rural areas from farmer participation in GrowNYC's Greenmarkets (GM) in New York City based on primary data collected from farm vendors and market customers, and compare it to county-level human capital stocks derived from secondary data.[2]

Delphi method

The DM is a popular, qualitative forecasting tool that has been applied across a variety of disciplines since its conception in the early 1950s (Gupta & Clarke, 1996). The notion behind the DM is that the aggregate of a group will provide a forecast generally superior to that of most of the individuals within the group. The technique is largely applied to problems where historical data do not exist or when existing data are inappropriate (Rowe, Wright, & Bolger, 1991). In the context of food systems research, the DM has been applied to issues of food literacy (Vidgen & Gallegos, 2014), strategic planning (Rikkonen, Kaivo-oja, & Aakkula, 2006), market development (Padel & Midmore, 2005), food risks (Wentholt, Fischer, Rowe, Marvin, & Frewer, 2010), supply chains (Ilbery, Maye, Kneafsey, Jenkins, & Walkley, 2004), and market access (Henchion & McIntyre, 2005). The DM has not been used to forecast impacts of urban local food initiatives on rural community development.

Given the nature of the research problem, it is important to have input from a variety of research disciplines as well as extension and community stakeholders. Accordingly, two expert advisory panels were assembled. The research panel consisted of 16 interdisciplinary researchers from universities across the Northeast region of the United States representing a wide array of disciplines, including natural resources, veterinary medicine, planning, education, forestry, development sociology, sociology, agricultural economics, geography, human ecology, and food studies. The extension panel consisted of 25 stakeholders from the Northeast involved with food systems in some way, including government officials, planners, food distributors and processors, farmers market managers, agricultural lenders, nonprofits, and other funding organizations.[3]

The research panel met twice. Panelists brought with them to the first meeting written thoughts on the ways in which urban local food system initiatives may influence each form of community capital, along with whether the level of impact was likely to be high, medium, or low. The full panel then brainstormed and discussed potential impacts by type of capital and identified potential indicators. The panel subsequently divided into four groups, assigned two capitals each, and separately discussed and refined the earlier results. Presentations by the groups back to the full panel followed, and a set of prioritized impacts and potential indicators for each capital were agreed to collectively. The second meeting included a review of the first meeting's results, along with a discussion of the history and educational and outreach activities of GM led by its executive director. After the executive director left, the panel again split into four groups, assigned two capitals each, and separately identified specifically how the transmission of impact to rural communities would occur. Groups reported back to the full panel, and the processes of transmission were discussed and revised collectively.

The extension panel met once following the research panel meetings, partially as a check on the research panel results, but also to develop prioritized impacts and indicators themselves (the extension panel was not informed of the research panel's results). Given the panelists' familiarity with GM, at the start of the meeting each shared specific examples of how farmer participation in GM impacted the farmers and the rural communities in which they resided. The executive director of GM followed with a presentation on the history and educational and outreach activities of GM, after which panelists were introduced to the RWC framework and engaged in discussion of where the examples initially provided would fit in or not. The panel subsequently divided into four groups, assigned two capitals each, and separately discussed and refined impacts and indicators. Presentations by the groups back to the full panel followed, in which a set of prioritized impacts and potential indicators for each capital were agreed to collectively.

Data collection

Outputs from the DM application were used as the basis for survey questions to farm vendors and customers at GM. The largest network of outdoor farmers markets in the US, in 2014 GMs existed in 54 locations and in all five boroughs of New York City (Figure 1). That year, 240 farmers and fishermen participated in GM representing six Northeast states (NY, NJ, PA, CT, VT, and MA), representing over 30,000 acres of land in production (Figure 2).[4]

Primary data – intellectual capital flows

Surveys to GM farm vendors were conducted through in-depth interviews from July 2015 through April 2016 that sell at one or more GM location ($N_F = 40$). Particular questions related

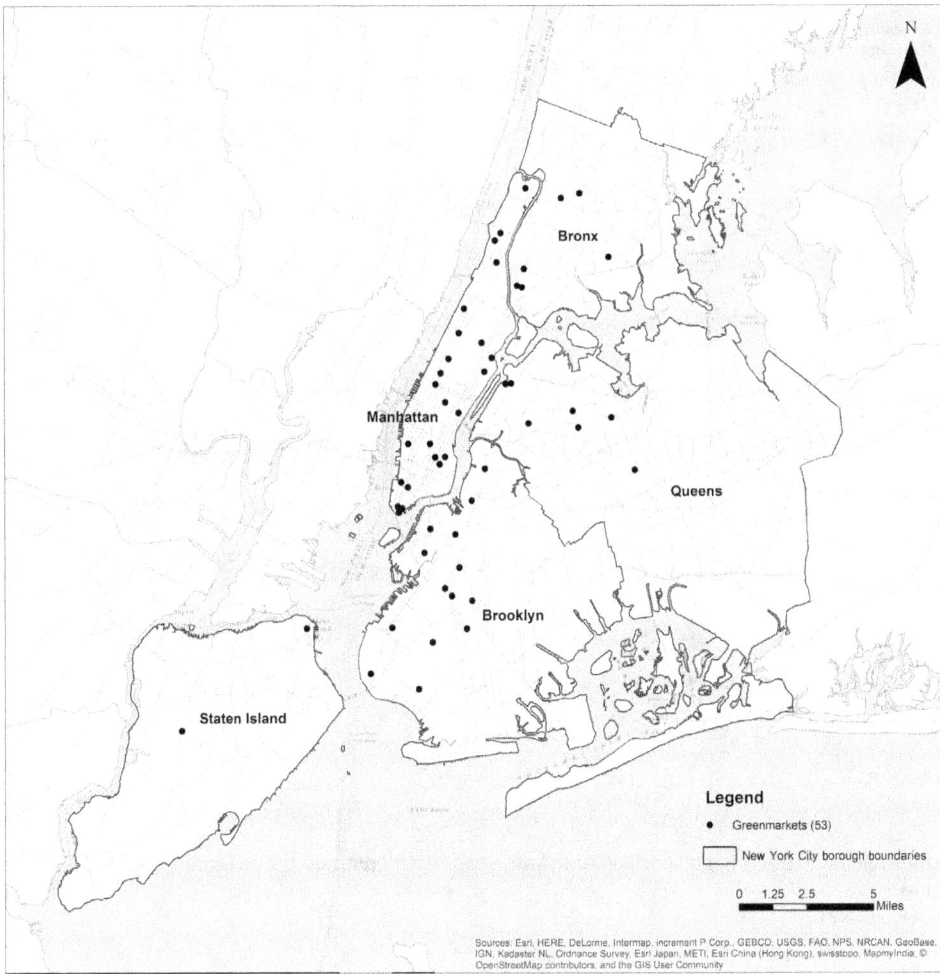

Figure 1. Map of Greenmarket locations, New York City, NY.
Source: GrowNY C.

to intellectual capital considered whether participation in GMs led to: (1) changes in production practices, the number of products and/or varieties grown, or the production of processed products; (2) development of new ideas for products and marketing techniques from interacting with other farm vendors, the GM manager, or GM customers, and whether those ideas were implemented at GM and at home (rural) markets; and (3) sharing of new ideas for products or marketing techniques learned from GM interactions with other farmers or individuals back in their home (rural) community.

Surveys to GM customers were administered in fall 2015 using Rapid Market Assessments (RMA), also known as dot poster surveys, at eight GM locations ($N_C = 824$). RMAs have many advantages, such as their ease of administration, the ability to get a large number of responses relatively quickly, and customer acceptance compared to written surveys or face-to-face interviews (Lev, Brewer, & Stephenson, 2008). However, questions must be limited in number and scope, and tracking the correlation of individual's responses across questions is infeasible. Following Thilmany McFadden et al. (2016), posters were refreshed regularly to

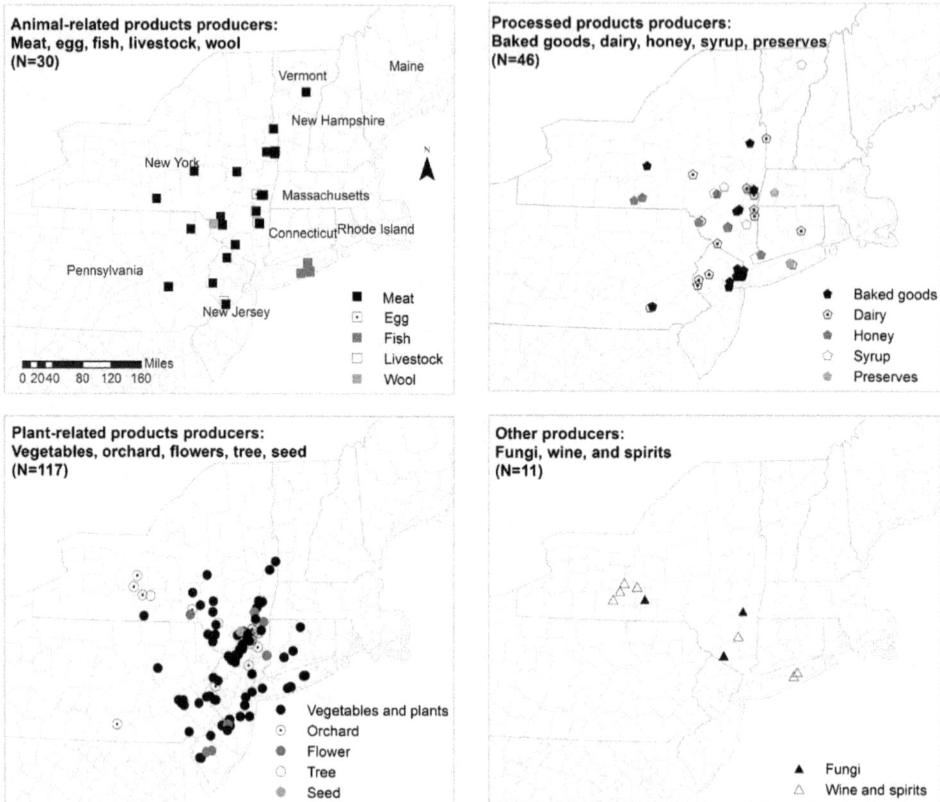

Figure 2. Map of Greenmarket participating farm vendors, by commodity and location, Northeast United States.
Source: GrowNY C.

reduce bias associated with respondents able to see customer answers that preceded them. Particular questions related to intellectual capital included: (1) whether customers talk to the farm vendors about what is happening on their farm, about policy issues (food, agricultural, and rural), and about ideas for new products, and (2) whether shopping at the GM influenced their overall perceptions about farmers, farms, agriculture, and rural places. Both questions relate to the types of information and knowledge exchanged between farmers and customers.

Secondary data – human capital stocks

Following Flora and Flora (2008), human capital comprises the skills and abilities of people and the ability to access outside resources and bodies of knowledge to increase understanding and identify promising practices. From Table 1, it is clear that both individual and intellectual capital fall within this definition. Accordingly, it is reasonable to combine the two definitions when utilizing secondary data variables that can fall under both definitions. A county-level index of the stock human capital was developed following the approach of Rupasingha et al. (2006) in their application to social capital. While the stock of human capital is not directly observable, it is a "structural condition that is expected to influence the covariance of a set of observed variables" (Rupasingha et al., 2006, p. 90). To create the index, a principal component analysis is conducted on the set of proxy variables and where the first principal component is interpreted as the index (or stock) of human capital.

Proxy variables were selected following Pender et al. (2012b), whether the data were available at the county level and consistent with application to local food systems. Six variables were included in the index (1) adult population proportion with a college-level degree, 5 year estimate, 2011–2015 (US Census Bureau, 2016); (2) labor force proportion in the younger-age cohort (25–44), 5-year estimate, 2011–2015 (US Census Bureau, 2016); (3) labor force participation proportion in nonfarm work, 5-year estimate, 2011–2015 (US Census Bureau, 2016); (4) labor force participation proportion in farm work, year 2012 (USDA, 2017); (5) proportion of farm operators with at least 10 years of experience, year 2012 (USDA, 2017); and (6) proportion of employment self-employed, year 2015 (US Department of Commerce, 2016).

Findings

The discussion of empirical findings begins with the DM application and is followed by the empirical outputs from the primary and secondary data collection efforts.

Impacts and indicators

Prioritized impacts for each form of community capital were harmonized across the research and extension advisory panel outputs (Table 1). For ease of exposition, much of our discussion will focus on the impacts for intellectual capital, namely: (1) market and industry education to and from urban and rural communities; (2) entrepreneurial innovation and idea sharing among farmers; and (3) product and value chain innovations in meeting consumer demand. However, there clearly exist potential for interaction between intellectual and other capital flows based on the impacts derived from the DM. By examining the impacts prioritized for all capitals, intellectual capital adaptations will likely influence or be influenced by supply chain infrastructure (built capital), farmer networks (social capital), land use decisions (natural capital), and connections to other institutional buyers in urban areas (political capital). If intellectual capital gains are further supported by the local food system initiative itself through technical assistance, business skills development, and farmer-customer networking opportunities, relationships to financial, individual, and cultural capital flows are evident.

More detail from the DM regarding intellectual capital follows in Table 2. While the extension panel results were consistently positive, the research panel results included cases for both positive and negative influences. In particular, negative influences consider whether infrastructural resources are sufficient to support product and value chain innovations (rural area), whether public education is resource constrained (in both rural and urban areas), whether existing GM rules limit the potential success innovations (i.e. restrictions on sources of products or types of ingredients), and whether sufficient communication with farmworkers exists to support and reinforce farm-level changes.

Ten indicators proposed from the DM application are shown in the last column of Table 2. As expected, the indicators largely focus on interactions between farmer vendors

Table 1. Capital definitions and potential rural impacts from farmer participation in urban-based farmers markets[a].

Type	Definition	Potential impacts
Built	Stock of fully functioning constructed infrastructure	• Expansion/innovation of supply chain infrastructure • Farmer investment in equipment and structures • Public investment in farmers market infrastructure
Financial	Stock of unencumbered monetary assets invested in other forms of capital or financial instruments	• Farm investment in land and infrastructure and/or increased utilization of existing farm assets • Market access expansion reduces farm financial risk • Urban market training and technical assistance improves grant success and productivity
Individual	Stock of skills and physical and mental healthiness of people	• Improved production and business skills for existing and new farmers • Farm labor opportunities and skills development • Health impacts from expanded sustainable agriculture practices • Physical and mental health improvement for participating farmers
Intellectual	Stock of knowledge, innovation, and creativity or imagination	• Market and industry education on consumer demands in large urban area • Increases rate of entrepreneurial innovation and idea sharing among farmers • Product and value chain innovations to meet or create consumer demand
Natural	Stock of unimpaired environmental assets (air, water, land, flora, fauna)	• Expanded agricultural land use • Soil health and water quality improvements from sustainable al production practices • Increased variety of plants and crops
Political	Stock of power and goodwill held by individuals, groups, or organizations that can be held, spent, or shared to achieve desired ends	• Increases in political conversations and support around agricultural policy issues from informed urban market customers • Increased ability of farmers to influence local, regional, and state policy from increased market/consumer demand information • Urban farmers markets improve connections and access to large/institutional urban market buyers.
Social	Stock of trust, relationships, and networks that support civil society	• New and/or expanded farmer networks • Social spaces for farmer-customer sharing • Civic engagement in rural areas
Cultural	Stock of practices that reflect values and identity rooted in place, class, and/or ethnicity	• Changes in production practices for more diverse consumer base • Improved market interactions between farmers and customers • Shifting identities of farmers and appreciation of urban customer base • Reintegration of farmers and customers across cultural barriers

[a]Capital definitions from WealthWorks (n.d.). Impacts summarized from Delphi Method application with research and extension advisory panels.

Table 2. Delphi method results on prioritized impacts of and proposed indicators for intellectual capital flows to rural areas from farmer participation in Greenmarkets (GM).

Prioritized impacts from advisory panels		
Extension panel	Research panel	Proposed indicators
GM educates people (farmers and consumers) that it is possible and cool to be a farmer, a career with a future, promoting rural youth retention in agriculture	Market and industry education to and from urban and rural communities • Demystification – of city for farmers, of farming for customers (+) • Increased knowledge of food system among consumers (+) • Increased knowledge for farmers of consumer demands (+) • Urban consumer experimentation with new products, new ideas (+/–) • Promotes youth education on cooking, agriculture, health (+) • Strain on rural human resources, expertise, capacity, competition (–) • Limit on public resources (cooperative extension, schools) to help facilitate innovations and new farmer training (–)	• Urban perceptions of agriculture and rural places • Urban understanding of policy issues related to agricultural and rural communities • Farmers better informed of consumer demands • Level of public education on agriculture
Marketing to GM leads to collective knowledge of opportunities and exploration of other and/or newer markets	Rate of entrepreneurial innovation and idea sharing among farmers • Increasing collaborative networks of farmers, idea sharing at GM (+) • Limited intellectual network expansion with rural (non-GM producers) (–) • Immediate feedback with a larger consumer audience at GM (+) • Increased knowledge of and stimulus to traditional/new production practices, new products, impacts on profitability (+/–) • Greenmarket rules may limit innovation (–)	• GM farmers share new ideas, marketing techniques with other GM farmers • GM farmers share new ideas, marketing techniques with rural area farmers • Change in farmer products, varieties, practices
GM formal and informal education leads to new kinds of value chain linkages and product development/ processing initiatives	Product and value chain innovations to meet or create consumer demand • Creative class connections (creating an environment in which entrepreneurial people want to live and work) or gentrification, rural redevelopment (+) • Promotes linkages with local supply chain intermediaries (+) • Misalignment with rural technical, infrastructure capacity (–) • Limited farmworker sharing of ideas about what is required (–)	• Farmers expand into processed products • Farmers increase linkages with downstream intermediaries • New or increased capacity of rural value chain infrastructure

and customers at the GMs and how they influence changes in urban perceptions about agricultural and rural issues, farmer idea sharing, and new product development or linkages with value chain intermediaries.

Greenmarkets-induced intellectual capital flows

Responses made by farm vendors and market customers provide important insight into forecasting how rural communities are impacted by investment in urban farmers markets. With respect to the farm respondents, 76% reported they got at least one new idea for new products and/or marketing techniques from selling at GM. Further still, of those that got at least one new idea, 77% had already made changes to their farm business based on these ideas (an additional 12% intended to make changes in the near-term). As an example, one vendor reported, "We would have never thought of selling garlic scapes, ramps, and some varieties of apples." Furthermore, 45% made their product and/or marketing changes in *both* urban and rural markets, and 65% shared ideas from GM with farmers or other individuals in their home communities (an additional 12% intended to do so in the near-term).

Of the changes made by vendors, increases in the interest and ability of farmers to produce and sell value-added (processed) products were prevalent. Sixty-five percent of respondents reported changes in this area as a direct result of their participation in GM. As an example, one vendor reported, "The kitchen [built on the farm, in my rural community] is completely because of Greenmarkets. Jams, fruit leather, baked goods including pies, etc. are what allow us to be at the market year-round."

Turning to GM customers, 49% of the RMA participants reported that they discuss what is happening on the vendor's farm when purchasing products, with 37% explicitly talking about opportunities for new products. Between 60 and 70% of consumers reported that shopping at GM has influenced their perceptions about farmers (69%), farms (61%), or agriculture (62%). Of those that reported it did not, most mentioned that they grew up on a farm or in a rural area. That said, only 30% of respondents indicated that shopping at GM and interacting with the farm vendors influenced their perceptions about *rural* places.

Stocks of human capital

Constructed from county-level data, the levels and variation in the stock of human capital are represented in Figure 3. As the index is the value of the principal component, the actual values of index by county (ranging from −5.37 to 2.44 in the Northeast) are less important than the values *relative* to one another. To interpret the figure, the darker the county, the larger is the estimated stock of human capital. Note that the stock of human capital is not directly affected by the magnitude of the population, but rather on the particular characteristics of the population via the proxy variables indicated above.

To relate the types of intellectual capital flows estimated from primary data to the estimated stocks of human capital, we compute average human capital indexes for six different multi-county areas in NYS: (1) all NYS counties, (2) all NYC counties, (3) all non-NYC counties, (4) non-NYC GM counties (counties within the GM mandated vendor region of 250 miles from NYC) that have GM vendors, (5) non-NYC GM counties with no GM vendors, and (6) non-NYC counties outside the GM mandated vendor region (Table 3). Further detail on the number of GM farm vendors residing in each county (NYC and non-NYC) provides additional richness to our analysis. Areas indicated with a "w" in Table 3 represent area statistics weighted by the number of GM vendors in each county.[5] Given the detailed orientation of the areas

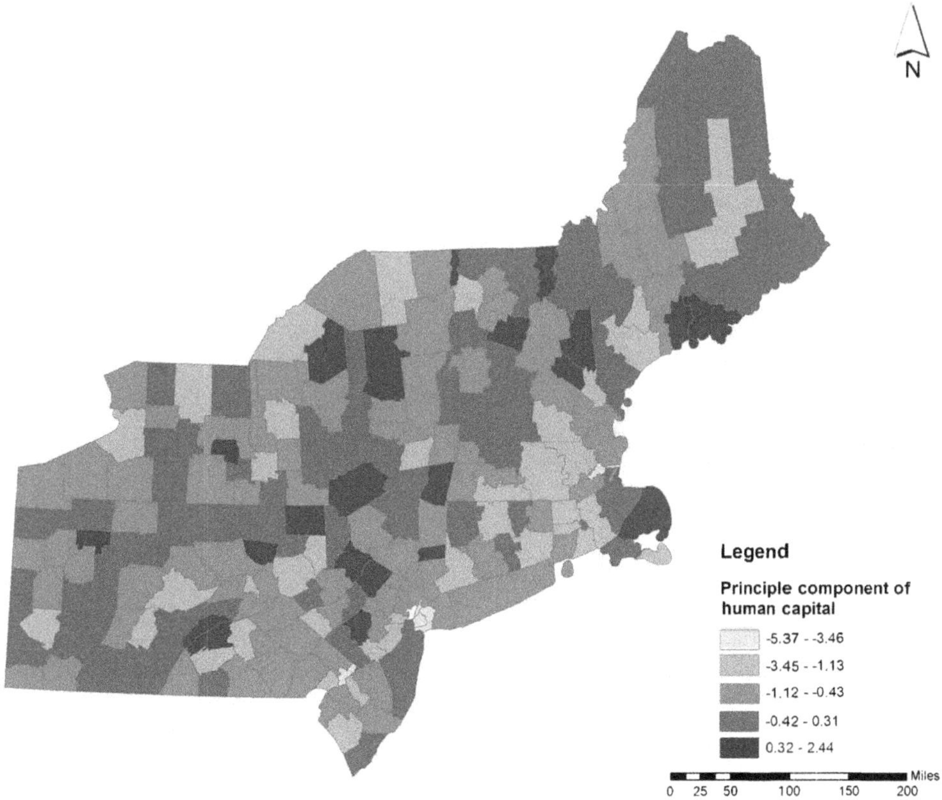

Figure 3. Author-derived index of human capital stocks, by county, Northeast United States, based on analysis of secondary data from 2011 to 2015.
Source: Author.

already, this may seem superfluous; however, using the weighted statistics allows us to explicitly account for differences in vendor concentrations. As one would expect, counties within the GM vendor region and closer to NYC tend to have more vendors residing there. Means-difference tests are used to compare whether the computed means are statistically different from one another. In particular, the null hypothesis is that the difference between the two means is zero, where the *p*-value represents the probability of obtaining the observed difference if the null hypothesis were true.

Ignoring GM for a moment, counties in the NYC area, on average, have lower stocks of human capital than counties outside of that area and by a relatively large margin (test 3 = 2: difference = 2.086, *p*-value < .001). Notably, the difference grows when limiting the non-NYC area to only counties in the GM region with GM vendors (test 4 = 2: difference = 2.268, *p*-value < .001), particularly when the concentration of GM vendors is factored in (test 4w = 2w: difference = 3.193, *p*-value < .001).

A more particular examination of human capital stocks in NYS counties considering GM farm vendor participation is available by comparing average stocks in the non-NYC GM areas for counties with and without GM farm vendors. In this case, the differences are smaller and nearly identical (in absolute terms) whether weighted or not (test 4 = 5: difference = .389, *p*-value = .052; test 4w = 5: difference = .391, *p*-value – .039), but only

Table 3. Descriptive statistics and means difference tests for human capital index, New York State.

Area no.	Area description	*N*	Average	Std. dev.	Minimum	Maximum
1	All counties	62	−.689	1.014	−4.706	1.168
2	NYC counties or NYC GM counties with GM vendors	5	−2.604	2.225	−4.706	−.018
2w	NYC GM counties with GM vendors (weighted)	5	−3.527	1.800	−4.706	−.018
3	NNYC counties	57	−.521	.641	−2.321	1.168
4	NNYC GM counties with GM vendors	27	−.336	.551	−1.548	.758
4w	NNYC GM counties with GM vendors (weighted)	27	−.334	.478	−1.548	.758
5	NNYC GM counties without GM vendors	19	−.725	.768	−2.321	1.168
6	NNYC NGM counties	11	−.623	.514	−1.394	.105
Test areas	Means difference test description (Null = 0)	df	Difference	Std. err.	*t*-stat	*p*-value
3 = 2	NNYC counties = NYC counties	60	2.086	.394	5.289	< .001
4 = 2	NNYC GM counties with GM vendors = NYC GM counties with GM vendors	30	2.268	.468	4.848	< .001
4w = 2w	NNYC GM counties with GM vendors = NYC GM counties with GM vendors (weighted)	30	3.193	.687	8.261	< .001
4 = 5	NNYC GM counties with GM vendors = NNYC GM counties without GM vendors	44	.389	.194	2.002	.052
4w = 5	NNYC GM counties with GM vendors = NNYC GM counties without GM vendors (weighted)	44	.391	.184	2.128	.039
4 = 6	NNYC GM counties with GM vendors = NNYC NGM counties	36	.286	.194	1.479	.148
4w = 6	NNYC GM counties with GM vendors = NNYC NGM counties (weighted)	36	.289	.175	1.651	.107

Notes: NYC = New York City, GM = Greenmarkets, NNYC counties = not NYC counties, NGM counties = not GM counties. NYC counties include Bronx, Kings, New York, Queens and Richmond. All NYC counties include at least 1 GM vendor. GM counties include all counties within 250 miles of NYC (in whole or in part). Weighted statistics are provided for areas with at least 1 GM vendor in each county. The weights are the counts of GM vendors, ranging from 1–5 to 1–26 for NYC and NNYC, respectively.

statistically different from zero when vendor concentration is considered. This is not too surprising given the closer proximity of the counties to one another compared with the previous tests.

The last empirical test considers human capital stocks in counties within the GM vendor region and with GM vendors relative to counties outside of the GM vendor region. In this case, the differences are not statistically different from zero, whether weighting or not (test 4 = 6: difference = .286, *p*-value = .148; test 4w = 6: difference = .289, *p*-value = .107).

Discussion

The DM applied to a very large and unique urban farmers market initiative supported the notion that net intellectual capital flows to rural areas were positive, particularly with market programming that includes strong educational efforts. The influences of multiple dynamics within and between urban and rural areas were identified and are summarized in Figure 4. The DM application formalized that, while GM provides market access for producers to a large consumer base, it also serves as a medium of and facilitator for vendor-to-customer and vendor-to-vendor interactions. The expert panel conclusions highlight conceptually that GM farm vendors bring enhanced levels of intellectual capital back to their home (rural) region, but that the ultimate impacts on the stock of intellectual capital (and thereby community outcomes) must be considered in the context of rural resource constraints (Figure 4).

From the urban perspective, the DM process concluded that extending knowledge gains by market customers on agricultural and rural issues to enhance public education efforts and improved political capital within the urban area is necessary before those benefits can translate to rural areas. The ability of this to happen in light of GM's overall footprint within a vast urban area is uncertain, as is the extent of political interest from urban-based elected officials in understanding the economic and political linkages between rural and urban regions (Figure 4). GM is the largest farmers market system in the United States; however, NYC is equally massive and unique. Ultimately, it is the relationships that matter.

Primary data from farm vendors and market customers provided evidence on the positive intellectual capital flows to rural areas. We found that GM was beneficial in improving urban perceptions about farming and agriculture; however, improving urban perceptions about rural places in general were much less apparent. Although the latter is equally important in informing rural development policy, a lack of a clear idea of what "rural" means, compared with a "farmer" or "agriculture," likely drives this result.

Primary data analysis revealed that GM farm vendors gained additional knowledge, new ideas, and improved skillsets to develop new products and varieties and to enhance their marketing techniques. By extending these knowledge gains to rural local food markets in which they participate, where applicable, gains to rural community outcomes should accrue. Additional rural gains are likely realized when GM vendors share these gains to producers in their home communities. We have shown empirically that participation in GM encourages the consideration and adaptation of additional supply chain activities for some producers, expanding their farm business into manufacturing value added food products and, importantly, increasing the length of their market season. To the degree that these activities are

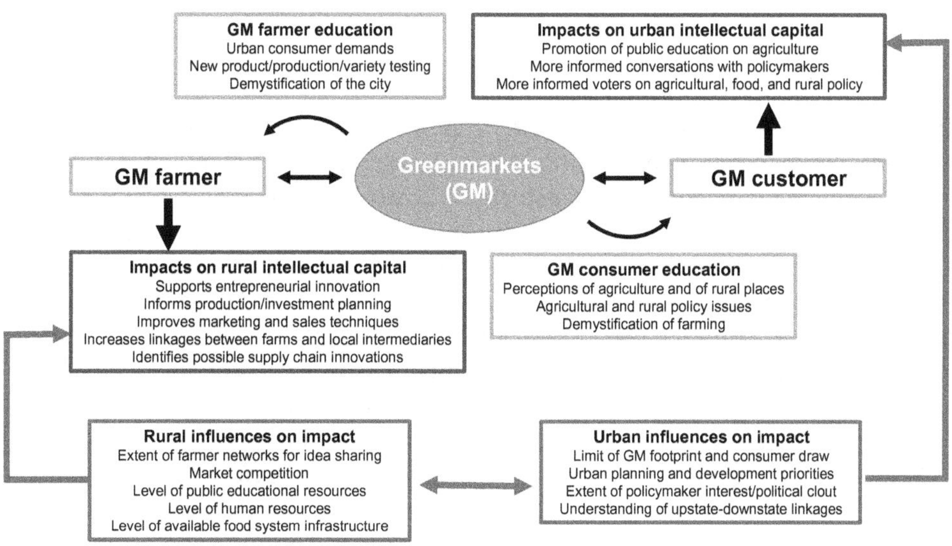

Figure 4. Conceptual representation of intellectual capital flows of Greenmarkets.

successful and expansive in the long term, they may support needed infrastructural and educational resource investments in rural areas.

Primary data analysis revealed positive intellectual capital flows to the producers participating in GM; that is to be expected if they are economically rational and the flows translate directly into farm benefits. Much less can be said from this, however, regarding more encompassing community outcomes. The additional farmer-to-farmer exchanges and supply chain activities for rural producers, as well as the increased availability of new products to rural consumers, represent positive flows, but are much more difficult to quantify. Accordingly, a county-level index for the stock of human capital was estimated from secondary data and evaluated based on the 51-county GM region (5 in NYC and 46 outside of NYC), the 11 counties not in the GM region, and the concentration of GM vendors in particular counties with means-difference tests. While arguably a simple test on the association of farmer participation in GM and the stock of human capital in farmers' home regions, it does provide empirical evidence on the issue before which none existed.

In general, we provide evidence that the stocks of human capital in NYC counties, on average, are lower than those in the counties that surround them. While this may be construed as an urban versus rural test, the simplicity of the test notwithstanding, there are several urban areas outside of NYC included in the non-NYC region that make this less clear. Given the small sample size, generalizations beyond this particular case are likely inappropriate. However, when we constrain our focus to GM counties outside of the NYC region and compare the counties with and without GM vendors, we find statistical evidence that human capital stocks are higher in the counties containing GM vendors when vendor concentration is considered. To examine the issue further, we considered human capital stocks in counties within the GM vendor region and with GM vendors relative to counties outside of GM vendor region. While anticipating spillover effects across counties within the GM region and considering the longer distance of the non-GM region from NYC, we were surprised the find no statistical difference in average stock levels across these two areas. However, this may be the result of the smaller size of the non-GM region with areas that include large urban areas, perhaps compatible with their own urban farmers market influences.

Conclusions

This study provides empirically based estimates of changes in intellectual capital flows to rural areas resulting from farmer participation in a large urban-based farmers market initiative in NYC. Furthermore, by comparing the stocks of human capital in counties with and without participating market vendors, we find that such participation in the urban market is associated with higher human capital stocks in the counties from which the vendors come. While appealing, generalizing these results to other urban farmers market initiatives is ill advised, given the size and scope of services this unique farmers market initiative provides. That said, to the degree that other studies follow the methodological approach developed here, valuable information will be gained by comparing the operational differences of the urban farmers market system with the differences in empirical outcomes.

Similar assessments with other capitals are currently in process for this GM case application, along with attention to potential interaction effects of alternative capitals stocks and flows. A clearer articulation of these capital relationships will be valuable in expanding the applications of these measures to other applications, such as an extended SAM to better compare different capital change measures, and how such capital accounting can be integrated into a comprehensive analysis linking these changes to other (traditional) economic sectors and impact measures.

Notes

1. Eight capitals have been described in the RWC literature, nearly identical to seven commonly used under CCF, but for a separation of human capital into individual and intellectual capital. The separation is not absolute nor overly important to this paper. Our initial attention on intellectual capital is the result of work with Yellow Wood Associates (WealthWorks, n.d.) and the advisory panel discussions where individual and intellectual capital were separately defined. However, as discussed later, our comparison to the stock of human capital is based on secondary data availability that is deemed relevant to both intellectual and individual capital flows.
2. GrowNYC was created in 1970 as the Council on the Environment of New York City. Originally a policy-based organization focused on quality of life issues like air quality, traffic, and noise, it established the city's first farmers market in several decades, the Greenmarket in 1976. Today, GrowNYC oversees many programs of which the Greenmarkets program is one.
3. Both panels were provided with definitions of the eight forms of capital (as defined in WealthWorks, n.d.) in advance of their first meeting.
4. The scale of GM and its customer base necessitates participation from vendors from a large geographic region that some may find disconnected from the definition "local." However, it is not uncommon for customers in the NYC area to consider the tristate area as "local" and, as Thilmany McFadden (2015) finds, localness varies by the audience, where geography is only one of the factors considered in its definition.
5. Inherently, this weighting scheme treats each GM vendor the same, i.e. weights are invariant to the size of the vendors' farms, length of time selling at GM, and the types and varieties of products they market.

Acknowledgements

The authors would also like to thank the members of the project's research and extension advisory teams for their participation and guiding contributions. Additionally, the project benefited from an experienced facilitator in the rural wealth creation framework, Shanna Ratner of Yellow Wood Associates, and from the research assistance of Jaclyn Hochreiter. Finally, the authors appreciate the participation of the entire GrowNYC/Greenmarkets team, and especially for the support from Michael Hurwitz.

Disclosure statement

The authors have no financial interest or benefit from the direct application of this research. The funders played no role in the case study design or the decision to submit the article for publication. The views expressed are the authors' and do not necessarily represent the policies or views of any sponsoring agencies.

Funding

This material is based upon work supported by the Agriculture and Food Research Initiative Competitive [grant number 2015-68006-22848] from the U.S. Department of Agriculture (USDA) National Institute of Food and Agriculture (NIFA); and the Farm Credit Northeast AgEnhancement Program [project number 75314/A001].

References

Arrow, K.J., Dasgupta, P., Goulder, L.H., Mumford, K.J., & Oleson, K. (2012). Sustainability and the measurement of wealth. *Environment and Development Economics, 17*, 317–353. doi:10.3386/w16599

Brown, J.P., Goetz, S.J., Ahearn, M.C., & Liang, C. (2014). Linkages between community-focused agriculture, farm sales, and regional growth. *Economic Development Quarterly, 28*, 5–16. doi:10.1177/0891242413506610

Brown, A., & Lewin, P. (2016, July). *Measuring United States county-level economic resilience to a recession*. Paper presented at the Agricultural and Applied Economics Association Conference, San Francisco, CA.

Deller, S.C., Brown, L., Haines, A., & Fortenbery, R. (2014). *Local foods and rural economic growth*. (University of Wisconsin-Madison Staff No. 570). Madison, WI: Department of Agricultural and Applied Economics.

Emery, M., & Flora, C. (2006). Spiraling-up: Mapping community transformation with community capitals framework. *Community Development, 37*, 19–35. doi:10.1080/15575330609490152

Fenichel, E.P., & Abbott, J.K. (2014). Natural capital: From metaphor to measurement. *Journal of the Association of Environmental and Resource Economists, 1*, 1–27. doi:10.1086/676034

Flora, C.B., & Flora, J. (2008). *Rural communities: Legacy and change* (3rd ed.). San Francisco, CA: Westview Press.

Goetz, S.J., & Rupasingha, A. (2006). Wal-Mart and social capital. *American Journal of Agricultural Economics, 88*, 1304–1310. doi:10.1111/j.1467-8276.2006.00949.x

Green, G.P., & Haines, A. (2016). *Asset building and community development* (4th ed.). Thousand Oaks, CA: Sage.

Gunter, A., & Thilmany, D. (2012). Economic implications of farm to school for a rural Colorado community. *Rural Connections, 6*, 13–16.

Gupta, U.G., & Clarke, R.E. (1996). Theory and applications of the Delphi technique: A bibliography (1975–1994). *Technological Forecasting and Social Change, 53*, 185–211. doi:10.1016/S0040-1625(96)00094-7

Henchion, M., & McIntyre, B. (2005). Market access and competitiveness issues for food SMEs in Europe's lagging rural regions (LRRs). *British Food Journal, 107*, 404–422. doi:10.1108/00070700510602183

Hughes, D.W., Brown, C., Miller, S., & McConnell, T. (2008). Evaluating the economic impact of farmers' markets using an opportunity cost framework. *Journal of Agricultural and Applied Economics, 40*, 253–265. doi:10.1017/S1074070800023580

Hughes, D.W., & Isengildina-Massa, O. (2015). The economic impact of farmers' markets and a state level locally grown campaign. *Food Policy, 54*, 78–84. doi:10.1016/j.foodpol.2015.05.001

Ilbery, B., Maye, D., Kneafsey, M., Jenkins, T., & Walkley, C. (2004). Forecasting food supply chain developments in lagging rural regions: Evidence from the UK. *Journal of Rural Studies, 20*, 331–344. doi:10.1016/j.jrurstud.2003.09.001

Jablonski, B.B.R. (2014a). Rural development through strengthened rural–urban linkages: The case of US local food systems. In P. Dannenberg & E. Kulke (Eds.), *Economic development in rural areas* (pp. 69–82). Burlington, VT: Ashgate.

Jablonski, B.B.R. (2014b). Wealth impacts of farmers' markets. In J.L. Pender, T.G. Johnson, B. Weber, & J.M. Fannin (Eds.), *Rural wealth creation* (pp. 218–231). New York, NY: Routledge.

Jablonski, B.B.R., Schmit, T.M., & Kay, D. (2016). Assessing the economic impacts of food hubs on regional economies: A framework that includes opportunity cost. *Agricultural and Resource Economics Review, 45*, 143–172. doi:10.1017/age.2016.9

Johnson, T.G., Raines, N., & Pender, J.L. (2014). Comprehensive wealth accounting. In J.L. Pender, T.G. Johnson, B. Weber, & J.M. Fannin (Eds.), *Rural wealth creation* (pp. 30–54). New York, NY: Routledge.

Kusmin, L. (2015). *Rural America at a glance, 2015 edition* (US Department of Agriculture Economic Research Service Economic Information Bulletin No. EIB-145). Retrieved from https://www.ers.usda.gov/publications/eib-economic-information-bulletin/eib-145.aspx

Lamie, R.D., Dunning, R., Bendfeldt, E., Lelekacs, J.M., Velandia, M., & Meyer, L. (2013). Local food systems in the South: A call for a collaborative approach to assessment. *Choices, 28*(4), 1–5.

Lev, L., Brewer, L., & Stephenson, G. (2008). *Tools for rapid market assessments* (Oregon State University Special Report 1088-E). Retrieved from https://smallfarms.oregonstate.edu/sites/default/files/small-farms-tech-report/eesc_1088-e.pdf.

Padel, S., & Midmore, P. (2005). The development of the European market for organic products: Insights from a Delphi study. *British Food Journal, 107*, 626–646. doi:10.1108/00070700510611011

Pender, J. (2015, July). *Measuring the comprehensive people-based wealth of rural communities*. Paper presented at the Agricultural and Applied Economics Association Conference, San Francisco, CA.

Pender, J., Marre, A., & Reeder, R. (2012a). Rural wealth creation: Concepts, measures, and strategies. *American Journal of Agricultural Economics, 94*, 535–541. doi:10.1093/ajae/aar076

Pender, J., Marre, A., & Reeder, R. (2012b). *Rural wealth creation: Concepts, strategies and measures* (US Department of Agriculture Economic Research Service No. ERR-131). Retrieved from https://www.ers.usda.gov/publications/pub-details/?pubid=44962

Pender, J.L., Johnson, T.G., Weber, B.A., & Fannin, J.M. (2014). Rural wealth creation: Introduction and overview. In J.L. Pender, B.A. Weber, T.G. Johnson, & J.M. Fannin (Eds.), *Rural wealth creation* (pp. 3–15). New York, NY: Routledge.

Pigg, K., Gasteyer, S., Martin, K., Keating, K., & Apaliyah, G.P. (2013). The community capitals framework: An empirical examination of internal relationships. *Community Development, 44*, 492–502. doi:10.1080/15575330.2013.814698

Putnam, R.D. (2007). E Pluribus Unum: Diversity and community in the twenty-first century The 2006 Johan Skytte prize lecture. *Scandinavian Political Studies, 30*, 137–174. doi:10.1111/j.1467-9477.2007.00176.x

Rikkonen, P., Kaivo-oja, J., & Aakkula, J. (2006). Delphi expert panels in the scenario-based strategic planning of agriculture. *Foresight, 8*, 66–81. doi:10.1108/14636680610647156

Rowe, G., Wright, G., & Bolger, F. (1991). Delphi: A reevaluation of research and theory. *Technological Forecasting and Social Change, 39*, 235–251. doi:10.1016/0040-1625(91)90039-I

Rupasingha, A., Goetz, S.J., & Freshwater, D. (2006). The production of social capital in US counties. *The Journal of Socio-Economics, 35*, 83–101. doi:10.1016/j.socec.2005.11.001

Schmit, T.M., Jablonski, B.B.R., & Mansury, Y. (2016). Assessing the economic impacts of local food system producers by scale: A case study from New York. *Economic Development Quarterly, 30*, 316–328. doi:10.1177/0891242416657156

Swenson, D. (2010). *Selected measures of the economic values of increased fruit and vegetable production and consumption in the Upper Midwest*. Ames: Iowa State University Leopold Center for Sustainable Agriculture Publication.

Thilmany, D., Bond, C.A., & Bond, J.K. (2008). Going local: Exploring consumer behavior and motivations for direct food purchases. *American Journal of Agricultural Economics, 90*, 1303–1309. doi:10.1111/j.1467-8276.2008.01221.x

Thilmany McFadden, D. (2015). What do we mean by "local foods"? *Choices*, 30(1), 1–6.

Thilmany McFadden, D., Conner, D., Deller, S., Hughes, D., Meter, K., Morales, A., … Tropp, D. (2016). *The economics of local food systems: A toolkit to guide community discussions, assessments and choices*. Washington, DC: US Department of Agriculture, Agricultural Marketing Service Report. Retrieved from https://www.ams.usda.gov/sites/default/files/media/Toolkit%20Designed%20FINAL%203-22-16.pdf

Trump, D. (2017). *Presidential executive order on promoting agriculture and rural prosperity in America*. Washington, DC: The White House, Office of the Press Secretary. Retrieved from https://www.whitehouse.gov/the-press-office/2017/04/25/presidential-executive-order-promoting-agriculture-and-rural-prosperity

United National Environmental Programme & Secretariat of the International Human Dimensions Programme on Global Environmental Change (UNU-IHDP and UNEP). (2012). *Inclusive wealth report 2012: Measuring progress toward sustainability (Summary for decision-makers)*. Bonn: UNU-IHDP. Retrieved from https://www.ihdp.unu.edu/docs/Publications/Secretariat/Reports/SDMs/IWR%20SDM%20Low%20Resolution.pdf

US Census Bureau. (2016). *2011–2015 ACS 5-year estimates. American community survey*. Retrieved from https://www.census.gov/programs-surveys/acs/technical-documentation/table-and-geography-changes/2015/5-year.html

US Department of Agriculture. (2015). *News Release* (Release No. 0142.15). Retrieved from https://www.usda.gov/wps/portal/usda/usdahome?contentid=2015/05/0142.xml

US Department of Agriculture, Census of Agriculture. (2017). *2012 Census of agriculture*. Retrieved from https://www.agcensus.usda.gov/Publications/2012/

US Department of Commerce. (2016). *Economic Profile CA30*. Bureau of economic analysis. Retrieved from https://www.bea.gov/iTable/download.fm?ext=pdf&fid=D9F0C015FF8FD727828024D5F0988FFEDFFC9CB7A023CD9C421591DD0E43766666C7D710C838006E97B45D8A2647D560348151DBC60DBDE7D9979B32461182C1s

Vidgen, H.A., & Gallegos, D. (2014). Defining food literacy and its components. *Appetite, 76*, 50–59. doi:10.1016/j.appet.2014.01.010

Vilsack, T. (2016, April). *New markets, new opportunities: Strengthening local food systems and organic agriculture*. Washington, DC: US Department of Agriculture. Retrieved from https://medium.com/usda-results/new-markets-new-opportunities-strengthening-local-food-systems-and-organic-agriculture-17b529c5ea90#.8lnvnyun7

WealthWorks. (n.d.) *Wealth: The eight capitals*. Retrieved from https://www.wealthworks.org/basics/explore-regional-wealth-building/wealth-eight-capitals.

Wentholt, M.T.A., Fischer, A.R.H., Rowe, G., Marvin, H.J.P., & Frewer, L.J. (2010). Effective identification and management of emerging food risks: Results of an international Delphi survey. *Food Control, 21*, 1731–1738. doi:10.1016/j.foodcont.2010.07.006

World Bank. (2006). *Where is the wealth of nations? Measuring capital for the 21st century*. Washington, DC: Author.

World Bank. (2011). *The changing wealth of nations. Measuring sustainable development in the new millennium*. Washington, DC: Author.

Food access, local foods, and community health

Steven Deller, Amber Canto, and Laura Brown

ABSTRACT
In this analysis, we look for patterns in the relationship between food access, local foods, and community health at the US county level. We take an ecological approach using 2012 Census of Agriculture and *County Health Rankings & Roadmaps* data collected by the University of Wisconsin Population Health Institute. We address a simple question: is access to local foods activity associated with public health? Results generally support the central premise, but the results are weak depending on the metrics of health and local foods activity chosen. In the end, we concluded that local foods are likely demand driven, and the mere presence of local foods activity is not sufficient to result in healthier community outcomes.

Introduction

Access to high-quality food, particularly within the context of the local food movement, emerged over the past decade as a combination of consumer movements promoting increased awareness of the environmental, geographic, and cultural implications of food production and consumption (Derkatch & Spoel, 2016; Guptill & Wilkins, 2002; Martinez et al., 2010). This movement is within the context that nearly 10% of the US population or 29.7 million people live in low-income areas more than one mile from a supermarket, and 12.7% of US households are food insecure (Coleman-Jensen, Rabbitt, Gregory, & Singh, 2016; Ver Ploeg, 2012). At the same time, non-communicable chronic disease – many of which are diet related (e.g. heart disease, stroke, cancer, and type 2 diabetes) remain the leading causes of death in the US (Heron, 2016). In some communities and neighborhoods, the development of local food systems is perceived as an opportunity to address these issues concurrently. As outlined by Story, Kaphingst, Robinson-O'Brien, and Glanz (2008), a major movement in the area of public health policy is to promote stronger links between local foods and consumers. Examples include the promotion of traditional farmers markets, farm-to-school programs, and as noted by Hardesty (2008), the coordination of institutional (e.g. hospitals, prisons, etc.) purchasing of local foods. Local food systems-based strategies are now being promoted at the community and neighborhood level to address a broad range of complex, systemic issues including environmental protection, restoring ecosystems, promoting sustainable economic development, or improving the health and wellbeing of local residents (e.g. Bimbo, Viscecchia, & Nardone, 2012; Feenstra, 1997, 2002; Minaker et al., 2014; Ruge & Mikkelsen, 2013; Webber, Stephenson, Mayes, & Stephenson, 2013).

In some cases, communities may view food systems strategies as an opportunity to simultaneously address multiple health, wellness, or agricultural and community vitality issues. As an example, Dunn (2013) refers to the concurrent farm income crisis and health crisis in terms of the "good food gap." That is, while farmers struggle to remain profitable in supplying the increased consumer demand for local foods, significant portions of the population either lack access to healthy foods or suffer from increasing rates of diet-related chronic disease. Similarly, Hamm (2008) argues that US agriculture production falls short of meeting the dietary recommendations established for public health promotion.

More recently, the provincial government of Ontario, Canada, passed Bill 36, the "Local Foods Act of 2013" to encourage the development of local food systems. As a result of this legislation, the Ontario Ministry of Health and Long-Term Care has embraced local foods as a mechanism to improve public health. In an analysis of the documentation and published materials around this effort, Derkatch and Spoel (2016) found that the Ministry asserted that: (1) local foods promote individual health; (2) local foods are good for the environment; (3) local foods support the local economy; and (4) local foods strengthen the community. They conclude, however, that "[a]s food studies scholars have emphatically shown over the past decade, many positive assumptions about local foods are romantic ideas rather than reflections of the reality of contemporary food systems, whether local or not" (p. 12). Rigorous scientific-based evidence is simply not sufficiently available at this time to support the claims made about benefits of local foods, and what evidence does exist suggests that the relationship between local foods and community outcomes is weak at best.

When it comes to public health, however, it is clear that access to nutritious food is a necessary but not sufficient condition (Morland & Evenson, 2009). Despite recent leveling off for decades, rates of both diet-related chronic disease, including obesity, and food insecurity increased substantially (Coleman-Jensen et al., 2016; Segal, Rayburn, & Martín, 2016). Nearly 38% of adults were obese in 2013–2014 (Flegal, Kruszon-Moran, Carroll, Fryar, & Ogden, 2016), and 17% of children aged 2–19 were obese in 2011–2014 (Ogden et al., 2016). Further, while the rate of growth of obesity prevalence appears to be leveling off (Segal et al., 2016), incidence and prevalence of type 2 diabetes continues to rise, with 9.3% of the population, or 29.1 million people in the US, with diagnosed diabetes (Centers for Disease Control & Prevention, 2014). In 2015, approximately 15.8 million households, or 12.7% of the US population, were classified as food insecure, meaning they lack assured access to affordable, nutritious foods at all times (Coleman-Jensen et al., 2016).

It is becoming widely argued that the complex nature of food systems issues requires innovative, systems focused, and transformative approaches (Clancy, 2013), including the promotion of local food systems (e.g. Creamer & Dunning, 2012; Garnett, 2013). Born and Purcell (2006) note that many local foods advocates assert that through the promotion of local food systems, a healthier society, social or economic justice, and/or environmental sustainability will result. Born and Purcell (2006) and more recently Derkatch and Spoel (2016) further note that the bulk of the available literature is less than satisfactory in terms of rigor and tends to be more speculative in nature (see Deller, Lamie, & Stickel, this issue). Indeed, few studies have attempted an empirical analysis of the relationships between characteristics of local food systems and public health outcomes (Albrecht et al., 2013). The analysis presented in this study aims to probe this research gap and follows the lead of Berning (2012), Salois (2012), Ferrer, Fonsah, Ramirez, and Escalante (2011) and to a less extent Minaker et al. (2014). Our primary research question is: do communities (proxied by US counties) with a higher concentration of characteristics associated with vibrant local food systems have positive public health outcomes?

A primary and necessary objective of our work was to define a measurement index for both local food systems and public health. Exploring and defining these indices is of interest to both researchers and practitioners seeking more rigorous methods for measuring and evaluating the effects of local food systems approaches and public health outcomes. Our methods for defining these measures are outlined in the methods section below. Our analysis makes use of 2012 Census of Agricultural data and population health data compiled by the University of Wisconsin Population Health Institute in collaboration with Robert Wood Johnson Foundation and 2010 Census data.

Beyond these brief introductory comments, the study is composed of the following additional sections: (1) a review of literature regarding agricultural characteristics and public health outcomes related to local food systems, (2) a description of our modeling approach and specification of methods including an outline of our measures of local foods and public health, (3) the modeling results, and (4) conclusions and policy implications noting the limitations of the work, and future research directions.

Literature review

Our literature review focuses on understanding known contributors to public health as well as the relationship between elements of local food systems and public health outcomes. This includes contributors to public health outcomes, which need to be controlled for in our analysis, influences on dietary behavior, understanding the food environment as a component of public health, the limited existing research on links between local food systems and public health, and known agricultural activities associated with local food systems. In essence, though we acknowledge that many characteristics, such as poverty and access to health care, influence public health, for the purposes of this analysis, we focus on food-related elements.

A number of food environment measures have been developed to expand the understanding of the relationship between health, diet, and disease, and the "local" food environment (Caspi, Sorensen, Subramanian, & Kawachi, 2012; Glanz, Sallis, Saelens, & Frank, 2005; Penchansky & Thomas, 1981). Caspi et al. (2012) conducted a systematic review of "local" food environment based on the notions of availability (adequacy of healthy food), accessibility (location of food supply and ease of getting to that location), affordability (food prices and personal perception of worth), acceptability (people's attitudes about attributes of the food environment), and accommodation (how well food sources accept and adapt to local residents' needs). Based on this expanded understanding of the food environment, we suggest that agricultural activities related to local foods might be conceptualized as a component of the food environment in that they can be defined in terms of geographic factors, production (fruits and vegetables vs. commodities), market (direct to consumer, institutional) characteristics, and economic factors.

The research is clear that diets high in fruits and vegetables are associated with a decreased risk of chronic disease (Fung et al., 2008; He, Nowson, & MacGregor, 2006; Hu, 2003; Montonen, Knekt, Järvinen, & Reunanen, 2004) and replacing higher energy density foods in the diet with foods of lower energy density, such as fruits and vegetables, can be an effective weight management strategy (Rolls, Ello-Martin, & Tohill, 2004; Tohill, Seymour, Serdula, Kettel-Khan, & Rolls, 2004). Strategies to increase fruit and vegetable consumption are increasingly linked to improvements in the food environment, particularly those focused on enhancing the availability of healthy food choices (Glanz et al., 2005; Kelly, Flood, & Yeatman, 2011; Story et al., 2008). Further, a growing number of strategies to improve "local" food environments are tied to local and regional food systems, and include farm-to-institution or farm-to-where-you-are programs,[1] farmers markets, community supported agriculture (CSA), and community and home gardens. The underlying rationale is that the direct connection with farmers enhances the consumer perception of the food and willingness of the participant to eat local and regional fruits and vegetables (Centers for Disease Control & Prevention, 2011). Despite the growing interest in local food and agriculture strategies, the inclusion of these initiatives in the literature describing "local" food environments has been limited.

The WIC Farmers Market Nutrition Program has arguably been most heavily studied for its influence on fruit and vegetable intake within the context of local foods (McCormack, Laska, Larson, & Story, 2010), with several studies showing increased intake of fruits and vegetables among program participants (Racine, Smith Vaughn, & Laditka, 2010) compared to non-participants (Herman, Harrison, Afifi, & Jenks, 2008; Kropf, Holben, Holcomb, & Anderson, 2007). Recent evaluation efforts of farmers market-based double value incentive programs for SNAP users, however, showed significant changes in shopping behavior among program participants, resulting in increased fruit and vegetable consumption while using purchasing incentives (Community Science, 2013; Lindsay et al., 2013; Young et al., 2013). In one evaluation study, the increases in fruit and vegetable consumption remained after the program ended (Wholesome Wave, 2012).

Farm-to-school programs may include varying degrees of connectedness with local produce, including local food procurement and inclusion in meal programs, field trips to local farms, agriculture education, nutrition education, and experiential learning such as garden programs. Joshi, Azuma, and Feenstra (2008) conducted an assessment of farm–to-school programs and found that 10 of 11 non-peer-reviewed evaluation studies confirmed improvements in dietary behavior when fresh, local produce is served in schools. The authors, along with others (Taylor & Johnson, 2013), however, call for additional standardized and more robust evaluation efforts.

To the best of our knowledge, few studies have looked at the impact of local food systems characteristics on diet-related disease and health (e.g. Berning, 2012; Ferrer et al., 2011; Jilcott et al., 2011; Minaker et al., 2014; Salois, 2012; Seguin et al., 2017). Berning (2012) used data from over 15,000 individuals in the New England states and matched them to the presence of farmers markets, CSA farms, and direct sales from farmers to people within their community. Berning's specific interest was to see if higher concentrations of local foods activity influenced individual weight outcomes. After controlling for individual characteristics, such as age, education, and gender, along with health characteristics including cholesterol and diabetes, Berning finds that the data generally support the central hypothesis. The relation, however, between CSAs and weight outcomes is less stable from a statistical perspective.

The overall results led Berning to conclude "[t]hese results provide evidence that local foods access can have potential indirect benefits" on weight outcomes (p. 57).

Salois (2012) examined the association of the built food environment and "local agriculture" with US county-level prevalence of obesity and diabetes. Total per capita dollar volume of direct farm sales was inversely associated with diabetes and obesity. Specifically, a $100 increase in per capita sales was associated with a .80% reduction in obesity rate and a 1.2% lower rate of diabetes. The density of farmers markets was also inversely related to the diabetes rate (Salois, 2012). Jilcott et al. (2011) examined associations among obesity and per capita farmers markets, grocery stores/supermarkets, and supercenters in US counties. They found that per capita farmers markets were inversely associated with the obesity rate in non-metro counties.

Ferrer et al. (2011) used a similar approach to Jilcott et al. (2011) based on US county-level data to examine the relationship between the concentration of farmers markets and direct sales from farmers to consumers on obesity and diabetes rates. Their results are consistent with the other handful of studies: the data generally support the expected relationship (a higher concentration of local foods activity is associated with better health outcomes), but the results are weak. The relative weakness of the results leads them to conclude that "[t]here is still no clear indication of its substantial impacts" (p. 11). All four of these studies reach the same general conclusion: higher levels of local foods activity are linked to better health outcomes but the strength and causation of that linkage is questionable. Our study attempts to build off of prior research by including a more expansive analysis of the food environment by exploring the relationship of county food systems activity with county-based diet-related health factors and health outcomes.

In a study of 2228 households Minaker et al. (2014) find, after controlling for age, gender, education, and income, that shopping at farmers markets or participation in CSAs tends to be associated with lower body mass and waist circumference. Seguin et al. (2017) found that the promotion of CSA interventions had a positive impact on the prevention of childhood obesity in low-income families. Participation in CSAs was found to increase awareness of nutritional values of fresh foods. This latter result would be consistent with the linkages between farm-to-school programs and increased understanding of food, nutrition, and health outcomes.

Modeling approach and methods

One of the difficulties in modeling health outcomes is that theory suggests that "everything matters" and there are numerous ways to empirically measure individual factors. From an ecological community level, for example, links between poverty and poor health outcomes are numerous, complex, and, as expressed by Murray (2006, p. 923) "inextricably intertwined" (Mansfield & Novick, 2012; Montgomery, Kiely, & Pappas, 1996; Pappas, Queen, Hadden, & Fisher, 1993). While there are numerous ways to define and measure poverty, it is important to consider which metric is most consistent with the underlying relationships being observed. Alternatively, one must ask, is poverty one of the driving factors of health, or is poverty a symptom of another underlying cause? For example, the notion of social capital, which is often tied to poverty, is gaining wider interest in the health literature (e.g. Cattell, 2001; Mackenbach et al., 2016; Szreter & Woolcock, 2004; Ziersch, Baum, MacDougall, & Putland,

2005). If, for instance, poor health is a symptom of poverty, and elements of social capital are the true underlying causes, the policy implications are significantly different.

These complications create problems for anyone interested in better understanding the relationship between local food systems and community health outcomes. In the broader sense, understanding community outcomes requires a systems thinking approach, such as the Shaffer Star (Shaffer, Deller, & Marcouiller, 2006) or the Community Capitals framework (Emery & Flora, 2006). These systems, unfortunately, do not lend themselves to comprehensive empirical analysis, and therefore the best that we can do is to pick pieces of the systems and test if the data support that particular piece of the system. In the end, the best that we can do is to acknowledge these complicating factors and be conservative in our policy interpretation of the empirical results.

Our basic model is built on a classical linear model where public health is a function of local foods and a series of other community characteristics. This is the same approach used by Salois (2012), Ferrer et al. (2011), and Jilcott et al. (2011) but differs from Berning (2012) and Minaker et al. (2014) in that they used health outcomes of individuals as opposed to using community wide health metrics. The model can be expressed as:

$$\text{PH} = \alpha + \beta\text{LF} + \gamma\text{FA} + \theta\text{HB} + \delta\text{EN} + \varphi\text{SE} + \vartheta\text{SC} + \pi\text{AF} + \tau\text{HC} + \varepsilon.$$

where PH represents our public health measures, LF is our measures of local foods, FA is access to food (other than local foods), HB is a set of health behaviors, EN is a measure of environmental health, SE is a collection of socioeconomic characteristics, SC is a set of social capital metrics, AF is a set of metrics to reflect access to physical activity opportunities, and HC is access to health care. Most of this data are drawn from the US Department of Agriculture Food Environment Atlas, the University of Wisconsin Population Health Institute's *County Health Rankings & Roadmaps*, the 2012 Census of Agriculture, particularly the USDA *Food Environment Atlas*, the Bureau of Economic Analysis Regional Economic Information System, County Business Patterns, and the American Community Survey.[2] Again, all data are for US counties and are reflective of the community in 2012.

Measuring public health

Given the complexity of public health issues and limitations of available data, developing a comprehensive metric of public health offers a challenge. This study builds on the work of the University of Wisconsin Population Health Institute who, in partnership with the Robert Wood Johnson Foundation, developed the *County Health Rankings and Roadmaps* program which allows for the construction of detailed population health profiles for every county in the US (University of Wisconsin Population Health Institute, 2013). Following Berning (2012), Salois (2012), Ferrer et al. (2011), and Jilcott et al. (2011), we use the percent of adults that are obese and the percent of adults that are diabetic. We also use a measure of premature mortality that is the years of potential life lost before age 75 (Vila, Booske, & Remington, 2006). Every death occurring before the age of 75 contributes to the total number of years of potential life lost. For example, a person dying at age 25 contributes 50 years of life lost, whereas a person who dies at age 65 contributes 10 years of life lost to a county's premature mortality. The premature mortality measure is presented as a rate per 100,000 population and is age-adjusted to the 2000 US population.[3] Higher values of each of these metrics of public health are associated with poorer overall levels of health. One of the limitations to the public health data is that the data are not available for all counties for every year. This means that data values may be missing for smaller, more rural counties. The selection of these particular metrics could be viewed as arbitrary as minimizing the problem of missing data was a consideration.

In addition to modeling each of these three health metrics independently, we also use principal components analysis (based on the correlation metric) to build a single scalar health index.[4] The idea is to use similarities in how the three health measures move together to build a single index.[5] The final principal component index accounts for 75.9% of the variation in the three individual health measures. A high level of explanatory power is intuitive because one would expect these three measures to be highly correlated. The individual weighting scheme, which is directive from the eigenvectors of the correlation matrix, is .5772 for obesity, .6008 for diabetes, and .5530 for premature mortality. Thus, higher values of all four measures are associated with poorer public health. This is important in interpreting the statistical modeling results below.

Measuring local foods

There is significant variation in terms of how local foods are defined in the vast food systems literature (Deller, Lamie, & Stickel, this issue; Goodman, 2013; McFadden, 2015; Sundbo, 2013). This has led to difficulties in both defining and measuring local food systems (Martinez et al., 2010). Durham, King, and Roheim (2009), for example, found that many consumers disagree that food produced beyond a 100-mile designation could be considered local food. In a study of food retailers, Dunne, Chambers, Giombolini, and Schlegel (2010) found that grocers' perceptions of local food varied significantly from one another. Martinez et al. (2010) find that opinions of what defines local also varied by the agricultural product being considered. In our review of existing literature, we found that local foods is characterized from the consumer or intermediated consumer perspective (Dunne et al., 2010; Zepeda & Li, 2006), in terms of proximity- distance, drive time, food-miles (Darby, Batte, Ernst, & Roe, 2008; Dunne et al., 2010; King et al., 2010; Zepeda & Li, 2006), by geo-political boundaries such as states (Darby et al., 2008), by ownership structure (local or non-local) of farm (Low & Vogel, 2011), relationship to place (Marsden, Banks, & Bristow, 2000), production techniques used, marketing channels used, size/scale (Low & Vogel, 2011), products grown, quality relationships to producers (King et al., 2010), or integration of the related supply chain (Marsden et al., 2000).

A second limiting factor is the availability of data at the community level. Farmers active in local food systems have several primary markets: (1) direct to consumers via farmers markets, road side stands, U-pick, or CSA type operations; (2) direct to retail grocery stores or restaurants; and (3) direct to institutions such as schools or prisons. While there is reasonable data on the number of farms within each distribution channel at the county level, there is limited information regarding the dollar value (a measure of size) of these channels. The recently published (2015) USDA study entitled "Local Food Marketing Practices Survey" makes some of this information available at the state and regional level, but these data are insufficient for the county level analysis we provide in this study.[6] Thus, our measures of local foods activity are capturing only a portion of the local foods system.

We build on what others (Berning, 2012; Ferrer et al., 2011; Jilcott et al., 2011; Salois, 2012) have used to study the relationship between local foods and public health and use the concentration of farmers markets within the community (county), the concentration of farms with direct sales to consumers, and the concentration of the number of CSA farms. Here we take the number of each type and adjust it by the community population (in terms of 100,000 persons). We also used direct sales per capita as a measure of the dollar volume of activity. Finally, unlike other studies, we also use the concentration of farms that reported direct sales to retailers. This latter measure attempts to capture intermediate sales, the second market identified by Low and Vogel (2011). All of these data are taken from the 2012 Census of Agriculture, except for the farmers market data that are compiled by the Agricultural Marketing Service and is available at the USDA *Food Environment Atlas*.

As with the public health measures, we explore each of these local foods metrics independently and as a scalar index. Using principal components analysis, the final index accounts for 53.1% of the total variation in the five metrics. While this is not as strong as the health index, generally as more variables are entered into a principal component, more "noise" is introduced into the analysis resulting in lower total explanatory power. The individual weights are .2849 for farmers markets, .4974 for concentration of farms with direct sales, .4670 for direct sales per capita, .4362 for concentration of CSA farms, and finally .5129 for concentration of farms with direct sales to retailers. While one generally thinks of farmers markets as the focal point of local foods, the location of the farmers market may not coincide with the location of actual farm production. Farmers markets tend to be located in more urban settings where consumers are concentrated and the farms themselves are more geographically scattered in more rural settings. Thus, our index of local foods should be considered more focused on the production or supply side than the market or demand side of the local foods system.

Control variables

As noted in the introductory comments of the methods section of this study, one of the difficulties in modeling public health is that theory suggests that "everything matters" and we have multiple ways of measuring each element. Thus, we build our set of control variables based on the tendencies in the existing literature with the acknowledgment that our final model may suffer from some misspecification. Our control variables reflect access to non-local foods, health behaviors that influence health outcomes, the local environment, key socioeconomic characteristics, social capital, access to physical activity opportunities, and access to health care.

Building off Walker, Keane, and Burke (2010), Morland and Evenson (2009), and Caspi et al. (2012), we use the concentration of grocery stores, supercenters and club stores, convenience stores, and specialized food stores as our measure of access to non-local food. We do not include the concentration of fast food or full-service restaurants in this particular analysis. Higher values of our measure are associated with greater access to food choices and hence improved public health. Health outcomes are also dependent on the behavior of individuals, such as smoking and excessive alcohol consumption (e.g. Brunello, Fort, Schneeweis, & Winter-Ebmer, 2016). Here we use data from the *County Health Rankings & Roadmaps* database and include the percent of adults that are smokers and the percent of traffic fatalities that involved alcohol as a proxy for excessive alcohol consumption. We expect higher levels of both smoking and alcohol related deaths to have a negative impact on public health outcomes. As noted by Brunekreef and Holgate (2002), the detrimental relationship between environmental pollution, particularly air pollution, and public health has been repeatedly documented within the scientific literature and thus should be controlled for in any study of public health. For this study, we use the fine particular matter (PM25) for a measure of air pollution. Ideally, we would prefer more detailed data but missing values for the most rural counties prevent us from using other measures of pollution.

The influence of socioeconomic characteristics of individuals on health has been well documented within the literature (e.g. Adler & Ostrove, 1999; Murray, 2006; Winkleby, Jatulis, Frank, & Fortmann, 1992). For this study, we control for the unemployment rate of the county (with the logic being that higher levels of unemployment creates stress which is detrimental to health), per capita income, the poverty rate, and the Gini Coefficient of income equality. Per capita income and the poverty rate move in opposite direction, and we would expect higher income (and concurrently lower poverty) to be linked to better health. The role of income inequality and health has also been extensively studied with the general conclusion that higher levels of inequality result in poorer health outcomes (e.g. Pickett & Wilkinson, 2015; Truesdale & Jencks, 2016). Thus, we control for such inequality with the Gini Coefficient where higher values of the Gini are associated with higher income inequality. Educational levels have also been directly linked to levels of public health where higher levels of education are associated with better health (Brunello et al., 2016; Stewart, 1995). For this study, we constructed an Education Index that captures the distribution of educational attainment for those age 25 years and over. Using the distribution of the population across seven educational attainment levels (i.e. less than a 9th grade education through graduate or professional degree), we compute the third moment of the distribution (skewness): if the Educational Index is negative, then the distribution of educational attainment is skewed toward higher levels and a positive value means that the distribution is skewed toward lower levels of education. Here we would expect a positive relationship between our Education Index and health outcomes.

We also include the percent of the population that is non-white as a control variable because there is sufficient evidence in the literature that communities dominated by populations of color tend to have poorer levels of public health (Abraido-Lanza, Chao, & Florez, 2005). We also include a simple Gini–Simpson index (also known as the Gibbs–Martin index as well as the Blau Index) of ethnic diversity. In a recent study, Churchill, Ocloo, and Siawor-Robertson (2016) find consistent evidence that higher heterogeneity in ethnic diversity is tied to poorer health outcomes. While they offer a wide range of potential theoretical linkages, a detailed discussion here is beyond the scope of this study. A higher value of the ethnic diversity index is associated with a more heterogeneous (diverse) community, and lower values are associated with a more homogenous (monocentric) community. We would expect a positive relationship between this index and health levels: a more diverse community leads to poorer health. Finally, we include the percent of the employment in farming, forestry, and fishing as an industry, mining, and manufacturing. Health risk factors are generally associated with these particular industries.

Perhaps the one community factor that has received the most attention within the public health literature over the past decade has been the role of social capital in health outcomes (Cattell, 2001; Kawachi, Subramanian, & Kim, 2007; Szreter & Woolcock, 2004; Wulff, Donato, & Lurie, 2015; Ziersch et al., 2005).[7] Arguably, social capital (norms, trust, reciprocity, and networks) could provide a theoretical construct to help understand findings such as higher income inequality and ethnic diversity being tied to poor health. One could also contend from a policy perspective that investing in community level social capital could help improve health outcomes. To control for potential influences of social capital on public health outcomes, we follow the work of Deller and Deller (2010) and Keene and Deller (2015) and use three distinct measures of social capital. First, we use the concentration of individual and family support groups (number of organizations per 100,000 persons), the violent crime rate, and the index derived by Rupasingha, Goetz, and Freshwater (2006). The RGF Index uses principal components analysis to combine several factors generally associated with social capital into a single index. Factors such as the concentration of religious organizations, business organizations, and voter turnout rates, among others, are included in the index. Higher values of the RGF Index are associated with higher levels of social capital.

The final three blocks of control variables include the rurality of the community (county), access to health care, and access to physical activity opportunities. While the degree of rurality (percent of the population living in rural areas as defined by the US Census Bureau) can be associated with many of the community characteristics we control for, there is sufficient evidence that rural residents view health care differently than their urban counterparts (Casey, Call, & Klingner, 2001) to warrant controlling for it in our modeling. It is widely accepted in the public health literature that the lack of access to health care is detrimental to overall health (Crilly, Keefe, & Volpe, 2011). If a community does not have ready access to doctors, other health care professionals, and preventative health care services, it is difficult to maintain high levels of public health. To control for access to health care, we combine the number of hospitals, physician offices, dentist offices, outpatient care centers, home health care providers, and number of primary care providers into a single concentration index. As with many of our other measures, we place the number of these services available in the community on a per 100,000 person basis. Within the context of public parks, Bedimo-Rung, Mowen, and Cohen (2005) show that access to opportunities that promote physical activity, such as public parks, increases public health. To control for this community characteristic, we include the number of golf courses, ski facilities, and health and fitness clubs, again on a per 100,000 person basis. Clearly, this is not an exhaustive list of physical activity opportunities, such as bike and/or walking paths, or tennis courts, but it is sufficient to act as a control variable (Table 1).

Empirical results

Our analysis proceeds in three steps. First, we estimated a simple set of correlations between our local foods and health measures (Table 2), then a regression-based analysis using ordinary least squares focusing on our local foods index (Table 3), and then re-estimation of the regression model with each individual element of the local foods index (Table 4). For the regression-based analysis, we adjust the variance estimates for the presence of heteroscedasticity and report standardized regression coefficients for ease in comparing the relative magnitude of influences across the different variables. The trade-off we face when controlling for such a wide range of potential contributors to public health characteristics is the risk of multicollinearity. Thus, we tested for multicollinearity and find that the condition index for the models range from 90.57 for the models with just the local foods index (Table 3) to 97.06 for the models with the individual local foods measures (Table 4). These are sufficiently larger than the traditionally accepted threshold of 50 but not above the definitive level of 100. Thus, our models may be subject to collinearity problems. Regardless, an examination of

Table 1. Descriptive statistics and sources of data.

	Mean	Median	Std Dev	Minimum	Maximum	Kurtosis	Skewness	Source
Health index	.00	−.04	1.51	−5.35	6.65	.55	.10	County Health Rankings & Roadmaps: University of Wisconsin Population Health Institute
Percent adults obese	30.59	30.70	4.23	13.10	47.90	1.26	−.34	County Health Rankings & Roadmaps
Percent diabetics	10.79	11.00	2.25	3.00	19.00	.20	.19	County Health Rankings & Roadmaps
Premature mortality	382.89	377.50	112.69	.00	1032.10	2.59	−.14	County Health Rankings & Roadmaps
Local food index	.00	−.44	1.63	−1.62	16.17	16.80	3.12	USDA Food Environment Atlas
Concentrations of farmers markets	.05	.03	.09	.00	1.36	34.37	4.45	Census of Agriculture 2012
Concentration of farms with direct sales	136.27	100.97	133.37	.00	1580.46	14.96	2.72	Census of Agriculture 2012
Direct sales per capita	7.85	4.02	12.73	.00	203.91	51.64	5.54	Census of Agriculture 2012
Concentration of CSA farms	11.19	4.63	19.59	.00	353.36	56.79	5.51	Census of Agriculture 2012
Concentration of farms with sales to retailers	44.07	26.11	57.25	.00	754.88	25.11	3.73	Census of Agriculture 2012
Concentration of food stores	106.10	27.00	351.55	.00	10,871.00	339.46	14.31	County Business Patterns
Percent of population smokers	18.41	19.60	9.24	.00	51.10	.24	−.50	County Health Rankings & Roadmaps
Driving death related to alcohol	31.57	31.21	15.44	.00	100.00	2.48	.73	County Health Rankings & Roadmaps
Air pollution PM25	11.61	11.85	1.54	7.18	14.90	−.73	−.40	County Health Rankings & Roadmaps
Percent of population rural	59.37	60.20	30.89	.00	100.00	−1.13	−.15	American Community Survey, US Census
Unemployment rate	7.68	7.52	2.75	.82	28.27	2.25	.74	American Community Survey, US Census
Per capita income ($000)	37.56	35.32	10.62	15.69	185.03	19.69	2.83	American Community Survey, US Census
Gini coefficient of income equality	.44	.44	.03	.33	.65	1.10	.47	American Community Survey, US Census
Poverty rate	16.82	16.10	6.44	1.00	47.90	1.31	.86	American Community Survey, US Census
Education index	1.24	1.30	.61	−.61	2.55	−.40	−.41	American Community Survey, US Census
Ethnic diversity index	.28	.25	.18	.02	.75	−1.25	.33	American Community Survey, US Census
Percent of population non-white	21.30	13.97	19.35	1.19	96.74	1.01	1.25	American Community Survey, US Census
Percent of employment in farming, forestry, and fishing	8.77	6.89	7.41	.01	46.78	1.06	1.13	Buruea of Economic Analysis, Regional Economic Information System
Percent of employment in mining	2.33	.59	4.82	.00	78.48	36.48	4.71	Buruea of Economic Analysis, Regional Economic Information System
Percent of employment in manufacturing	8.74	6.97	6.84	.00	55.71	3.05	1.47	Buruea of Economic Analysis, Regional Economic Information System
Rupasingha, Goetz, & Freshwater social capital index	.00	−.23	1.33	−3.38	17.55	12.44	1.96	Rupasingha, Goetz & Freshwater
Concentration of individual and family support services	29.59	20.50	50.95	.00	1323.74	236.34	12.63	County Business Patterns
Violent crime rate	240.18	189.87	205.20	.00	1723.72	5.12	1.77	FBI Uniform Crime Reports
Access to health care	147.00	135.62	79.38	.00	1014.97	5.61	1.22	County Health Rankings & Roadmaps
Access to fitness activities	15.02	12.85	15.03	.00	421.35	188.32	8.54	County Health Rankings & Roadmaps

Table 2. Simple Pearson correlation coefficients between health and local foods measures.

	Percent adults obese	Percent diabetics	Premature mortality	Heath index	Concentration of farmers markets	Concentration of farms with direct sales	Direct sales per capita	Concentration of CSA farms	Concentration of farms with sales to retailers
Percent diabetics	.7154*** (.0001)								
Premature mortality	.5563*** (.0001)	.6416*** (.0001)							
Health index	.8711*** (.0001)	.9068*** (.0001)	.8346*** (.0001)						
Concentrations of farmers markets	−.0892*** (.0001)	−.1059*** (.0001)	−.1754*** (.0001)	−.1406*** (.0001)					
Concentration of farms with direct sales	−.1183*** (.0001)	−.1349*** (.0001)	−.2702*** (.0001)	−.1980*** (.0001)	.2849*** (.0001)				
Direct sales per capita	−.1472*** (.0001)	−.1752*** (.0001)	−.2288*** (.0001)	−.2099*** (.0001)	.2304*** (.0001)	.5741*** (.0001)			
Concentration of CSA farms	−.0814*** (.0001)	−.0657** (.0003)	−.0882*** (.0001)	−.0896*** (.0001)	.2429*** (.0001)	.3862*** (.0001)	.3609*** (.0001)		
Concentration of farms with sales to retailers	−.1552*** (.0001)	−.0941*** (.0001)	−.1768*** (.0001)	−.1616*** (.0001)	.2258*** (.0001)	.5946*** (.0001)	.5142*** (.0001)	.5684*** (.0001)	
Local food index	−.1646*** (.0001)	−.1571*** (.0001)	−.2581*** (.0001)	−.2201*** (.0001)	.4641*** (.0001)	.8104*** (.0001)	.7608*** (.0001)	.7106*** (.0001)	.8355*** (.0001)

Notes: Unit of analysis is U. counties for 2012.
Marginal significance, or *p*-values, in parentheses.
***Significant at the 99.9% level; **Significant at the 95.0% level; *Significant at the 90.0% level.

Table 3. Modeling results of local foods index on various measures of public health.

	Health index	Percent adults obese	Percent diabetics	Premature mortality
Local food index	−.1929 (.0168)***	−.1845 (.0517)***	−.1715 (.0263)***	−.1475 (1.4775)***
Concentration of food stores	−.0805 (.0001)***	−.1039 (.0003)***	−.0445 (.0001)**	−.0630 (.0057)**
Percent of population smokers	.1937 (.0024)***	.1848 (.0075)***	.1223 (.0037)***	.2028 (.2306)***
Driving death related to alcohol	.0099 (.0013)	.0112 (.0041)	−.0173 (.0020)	.0341 (.1255)**
Air pollution PM25	.0598 (.0138)***	.0454 (.0451)**	.0776 (.0226)***	.0315 (1.0795)**
Percent of population rural	.2305 (.0010)***	.1132 (.0034)***	.3243 (.0017)***	.1585 (.0849)***
Unemployment rate	−.0142 (.0100)	−.1329 (.0322)***	.0835 (.0160)***	.0094 (.8166)
Per capita income ($000)	−.0693 (.0034)**	−.0514 (.0126)	−.0621 (.0052)**	−.0680 (.2610)**
Gini coefficient of income equality	.0253 (.8691)	−.0426 (2.6436)**	.0841 (1.2900)***	.0222 (92.0190)
Poverty rate	.2806 (.0064)***	.2293 (.0196)***	.1648 (.0100)***	.3474 (.5827)***
Education index	.2645 (.0432)***	.2390 (.1444)***	.2845 (.0693)***	.1633 (3.2715)***
Ethnic diversity index	.0403 (.2692)	−.1068 (.8304)**	.1014 (.3742)**	.1113 (21.6511)**
Percent of population non-white	.1268 (.0030)**	.2485 (.0094)***	.0911 (.0042)**	−.0122 (.2352)
Percent of employment in farming, forestry, and fishing	.0035 (.0043)	.0619 (.0136)**	−.0174 (.0066)	−.0361 (.4181)
Percent of employment in mining	.0414 (.0051)**	−.0073 (.0137)	.0070 (.0076)	.1128 (.4913)***
Percent of employment in manufacturing	.0531 (.0026)***	.1165 (.0092)***	.0210 (.0046)	.0003 (.1938)
Rupasingha, Goetz, & Freshwater social capital index	.0102 (.0255)	.0522 (.0684)**	.0778 (.0363)**	−.1112 (2.6955)**
Concentration of individual and family support services	.0088 (.0003)	−.0073 (.0010)	−.0284 (.0004)**	.0626 (.0341)***
Violent crime rate	.0982 (.0001)***	.1172 (.0004)***	.0676 (.0002)***	.0722 (.0095)***
Access to health care	−.0264 (.0003)	−.1095 (.0012)***	.0109 (.0005)	.0306 (.0279)
Access to fitness activities	−.0333 (.0015)**	−.0192 (.0065)	−.0443 (.0032)**	−.0228 (.1977)
R^2	.6269	.4573	.5326	.5837
F statistic	240.39***	120.58***	163.05***	200.67***
n = 3027				

Notes: Unit of analysis is US counties for 2012.
Hetrosckedastic corrected standard errors in parentheses.
All reported coefficients are standardized.
***Significant at the 99.9% level; **Significant at the 95.0% level; *Significant at the 90.0% level.

Table 4. Modeling results of individual local food components on public health.

	Health index	Percent adults obese	Percent diabetics	Premature mortality
Concentrations of farmers markets	−.0417** (.2509)	−.0331** (.7691)	−.0397** (.3713)	−.0363 (30.2747)
	[−.2330 : −1.2165]	[−.0999 : −3.1147]	[−.2981 : −1.7535]	[12.3166 : −106.3601]
Concentration of farms with direct sales	−.1841*** (.0002)	−.1356*** (.0007)	−.1487*** (.0004)	−.1994*** (.0264)
	[−.0016: −.0025]	[−.0028: −.0057]	[−.0018: −.0032]	[−.1168: −.2202]
Direct sales per capita	−.0249* (.0018)	.0095 (.0068)	−.0512** (.0035)	−.0222 (.1828)
	[.0005 : −.0064]	[.0165 : −.0101]	[−.0023 : −.0158]	[.1618 : −.5546]
Concentration of CSA farms	.0296 (.0015)	.0169 (.0041)	.0094 (.0021)	.0531* (.1559)
	[.0053 : −.0007]	[.0117 : −.0044]	[.0051 : −.0030]	[.6109 : −.0002]
Concentration of farms with sales to retailers	−.0600** (.0006)	−.1166*** (.0018)	−.0196 (.0009)	−.0209 (.0620)
	[−.0004 : −.0027]	[−0.0051 : −.0121]	[.0010 : −.0026]	[.0805 : −.1626]
R^2	.6375	.4644	.5394	.5997
F statistic	211.13***	104.07***	140.58***	179.83***
$n = 3027$				
Control variables included	Yes	Yes	Yes	Yes

Notes: Unit of analysis is US counties for 2012.
Hetrosckedastic corrected standard errors in parentheses.
All reported coefficients are standardized.
Numbers in brackets is the 95% confidence intervals of the nonstandarized coefficients with the hetrosckedastic corrected standard errors.
***Significant at the 99.9% level; **Significant at the 95.0% level; *Significant at the 90.0% level.

the individual proportion variances reveals no consistent pattern in identifying troublesome variables. As a second test we also estimated the variance proportions, or variance inflation, for each variable, and none are above the critical value of 10. Three potential variables that might be troublesome in tandem are the poverty rate (VIF = 3.67), the ethnic diversity index (VIF = 4.21), and the percent of the population non-white (VIF = 5.05). Our findings hint at the presence of multicollinearity, but there is no definitive evidence. Given the relative stability of our results discussed below, we prefer to error on the side of including too many variables rather than too few, which might result in an overstatement of the influence of local foods on health.

Examining the simple correlation matrix of our local foods and health measures (Table 2), we find that the health and local foods indices move in the opposite direction, supporting our central hypothesis: higher levels of local foods-related activity are associated with lower levels of poor health outcomes. On its face value, this simple result is consistent with Berning (2012), Salois (2012), Ferrer et al. (2011), and Jilcott et al. (2011). The size of the correlation coefficient, however, is relatively small: −.22. A simple scatter plot (Figure 1) reveals that the expected relationship is present in the data, but it is a fairly weak relationship.[8] If we look at the individual local foods characteristics that make up our index and their simple correlation to our health index, we again find the expected negative relationships, each strongly significant from a statistical perspective, but weak from the size of the correlation coefficients. Looking across each individual local foods measure and health characteristics, we again find the expected inverse relationship that supports our hypothesis, but the size of the coefficients is small at best.

The fully estimated models provided in Tables 3 and 4 also generally support our hypothesis. In the first set of models where we look at the local foods index on the health index and

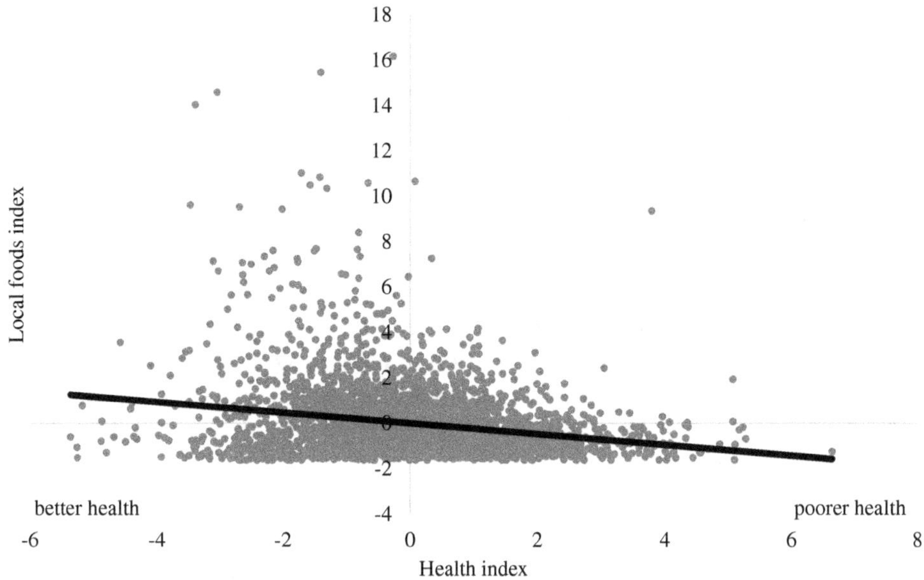

Figure 1. Simple scatterplot of local food and health indices for US counties, 2012.

its three separate components (Table 3), we find that, even after controlling for a wide range of characteristics that influence health, the statistically significant negative relationship between local foods and health remains. Across each health measure, higher levels of local foods activity are statistically linked to better health. Because we are reporting standardized coefficients we can directly compare and contrast the relative strengths of our local foods index and the four measures of health. The impacts on the health index, obesity, and diabetes are for all practical purposes the same, and lowest on premature mortality.[9]

Now consider the individual elements of the local food index (Table 4).[10] Here the data tend to support our central hypothesis, but the relationship is more nuanced. The concentration of farmers markets is linked to lower levels of obesity and diabetes, but only weakly to lower premature mortality. The concentration of farms with direct sale to consumers has the expected relationship across all health measures, but direct sales per capita has a weaker relationship. This is surprising because one would think that actual consumption of local foods, assumed through the measurement of direct sales per capita, would be more of an influence on health than a simple count of farms with direct sales. We also find that the density of CSA farms is not linked to better health and may be linked to higher levels of premature mortality. This result is unexpected. We also find some evidence that sales to intermediary markets are also weakly linked to better health outcomes. Based on the standardized coefficients, it is clear that the density of farms with direct sales to consumers has the strongest effect on health outcomes over concentration of CSAs, farmers markets, or availability through retail markets. This latter result suggests that raw availability of local foods may be more relevant than the specific interface between the farmer and consumer.

Further examination of the results of control variables offers interesting insights into community health (Table 3). The majority of the control variables performed as expected. For example, a higher concentration of retail food stores, higher income levels, and lower poverty rates is tied to better health outcomes. At the same time, a greater proportion of smoking in the community results in poorer health outcomes and a higher concentration of air pollution. Higher levels of education are linked to better health outcomes. We also find that the more rural the county, the worse the health characteristics. This latter result is obtained after controlling for other community characteristics that might explain why rural is "different" from urban. So we cannot say that poorer rural health outcomes are due to lack of access to food stores or health care; something more fundamental is occurring. One possible explanation is that Casey et al.'s (2001) observation that rural residents are less likely to seek out health care is at play. If we focus on the health index, the top four controls that have the largest impact include poverty rate (.2806), ruralness (.2305), smoking rate (.1937), and percent of the population non-white (.1268). Indeed, access to local foods has nearly same effect as percent of the population that are smokers, but in opposite directions (–.1929 vs. .1937). Only ruralness and poverty rates have a stronger affect than access to local foods as measured by our local food index.

Higher levels of employment shares in farming, forestry, fishing, as well as mining and manufacturing, tend to be linked to poorer health outcomes, but the results are somewhat inconsistent. For example, mining dependency is linked to higher levels of our health index, which means poorer health, and premature mortality, but is not associated with obesity or diabetes. We also find mixed relationships between the unemployment rate and health. Higher unemployment is linked to higher levels of diabetes but lower levels of obesity and no relation to premature mortality. It is possible that our simple measure of unemployment is not sufficient, and perhaps a measure on persistent unemployment might be a preferable measure. We also find mixed results with our ethnic diversity index. We find that more diverse communities tend to have lower obesity rates, but higher rates of diabetes and premature mortality. This inconsistency across the three individual health measures explains the lack of relationship between ethnic diversity and our broader health index. As expected, we do find that a higher share of the population that is non-white is associated with poorer health. A higher concentration of health care services is only linked to lower obesity rates, but not diabetes or premature mortality. One interpretation on the rural results may be that simple access to health care services is not sufficient to ensure better health, which is supported by the University of Wisconsin Population Health Institutes *County Health Rankings & Roadmaps* framework suggesting that access to health care accounts for only 20% of health outcomes (University of Wisconsin Population Health Institute, 2013). Finally, our simple measure of access to fitness activities has the expected relationship, but the statistical significance is mixed. Again, mere access to opportunities to be physically active may not be sufficient.

The three social capital measures provide somewhat contradictory results. For example, higher levels of violent crime are linked to higher levels of poor health across all three measures and our health index. This is not to suggest that violent crime causes poor health, but rather the lack of key characteristics associated with social capital tied to violent crime is also tied to health. The concentration of individual and family support services, however, is not linked to three of the four health measures and is actually associated with higher levels of premature mortality. This particular social capital measure may be insufficient, as it does not reflect the level of services offered through state and local governments. We also find mixed results with the most comprehensive social capital measure developed by Rupasingha, Goetz, and Freshwater. We find that higher levels of social capital are actually linked to higher obesity and diabetes rates but lower levels of premature mortality. This movement in opposite directions of the three separate health measures explains why the RGF Index is not statistically significant in the health index model. The inconsistency of the social capital results points to the difficulty in studying social capital from an empirical perspective.

Discussion

Local foods have been promoted as a means to reinvigorate rural communities through creation of economic opportunities and building local community capacity, enhance environmental sustainability, and promote better health (e.g. Lamie et al., 2013). The academic literature, however, has found it difficult to find rigorous evidence supporting these claims. The lack of rigorous evidence led Derkatch and Spoel (2016, p. 12) to conclude that "[a]s food studies scholars have emphatically shown over the past decade, many positive assumptions about local foods are romantic ideas rather than reflections of the reality of contemporary food systems, whether local or not." This statement largely holds true within the public health sector, as we drew from the available studies examining the relationship in public health outcomes and local foods activities foods (Berning, 2012; Ferrer et al., 2011; Jilcott et al., 2011; Minaker et al., 2014; Salois, 2012).

Our results examining the relationship between local foods activities and public health outcomes are largely consistent with the limited empirical literature. The data support the central premise and the results are statistically strong, but our results suggest that the promotion of local foods will have a modest positive impact on health outcomes. This is consistent with the findings of Jablonski, McFadden, and Colpaart (2016). In a study of Pueblo, Colorado, Jablonski and colleagues find that proactively improving the access to food environment may help enhance food security for some, but not all, subgroups. In a study of rural US communities, Deller, Canto, and Brown (2015) find that the breakdown between the access to healthy foods, again including local foods, is magnified through severe poverty. Our analysis suggests that the optimism about the relationship between the promotion of local foods and public health advocated by Salois (2012) may be exaggerated.

Individual consumer preferences drive decisions about whether or not to take advantage of access to healthy foods, including but not limited to locally sourced foods. Berning (2012) correctly observes, that in the larger market for food, the actual dollar value of local foods being purchased is small. Limiting ourselves to direct sales to consumers, the single best measure of the volume of local foods, in 2012 for the typical community (county) in our analysis, direct sales was only $7.85 per person with the largest level at $203.91. In 2013 the typical US household spent $6600 on food, or about $2640 per person. That means that the typical consumer spends less than one percent of their food dollars on local foods. Even for the community with the highest level of direct sales per capita, it is only 7.7% of food dollars.[11] Recent efforts to tie incentives to the use of federal food assistance programs (such as the Supplemental Nutrition Assistance Program, SNAP) at local food venues, such as farmers markets, have suggested there is potential to further build consumer demand and purchasing power of local foods for some individuals. Several studies of these incentive programs have documented increased revenue and sales at farmers markets, and have indicated the presence of the incentive program as a significant factor in determining where to spend food dollars (Community Science, 2013; Freedman, Mattison-Faye, Alia, Guest, & Hébert, 2014; Lindsay et al., 2013; Wholesome Wave, 2012; Young et al., 2013). The sustained impact of these programs on direct sales to consumers is yet to be determined. Regardless, using the logic of Berning (2012), even doubling the size of the local foods market is likely to result in very small changes in food purchasing patterns.

One of the underlying difficulties in documenting the potential impact of local foods on health, or any community outcome, is addressing causation. Is it possible that the demand for local foods is driving our results rather than the access to local foods itself? There is a small but growing literature on the demand for local foods (e.g. Zepeda & Li, 2006; Zepeda & Nie, 2012) that suggests that the consumer characteristics, such as education and income, are important in determining demand. These differences in demand for local foods explain the findings of Jablonski et al. (2016). Incentive programs are also attempting to shift consumer purchasing patterns, and arguably consumer demand. With regard to health outcomes, is it the case that the market for local foods is stronger in communities that are composed of healthier people? As noted by Webber et al. (2013), people who have the tendency to frequent farmers markets generally have a better understanding of nutrition than those that do not. It is possible, perhaps likely, that communities with a greater density of people who demand healthier foods may be associated with local production. Here, rather than the promotion of local foods as a means to promote health, it may be that healthier people demand locally sourced foods. If this is the case, the policy implications are strong: simply promoting local foods is not sufficient to promote health. Policies that take the approach of "build it and they will come" will fall short.

Conclusions

Many communities are looking at the promotion of local food systems as a means to improve their vitality. This can be through enhanced economic performance, stronger sense of community and social capital, and promoting better health outcomes. Unfortunately, as noted by Born and Purcell (2006) and Derkatch and Spoel (2016), among others, the research foundation upon which to justify the use of local foods is modest and often questions the validity of the strategies. In this study, we explore the relationship between local foods and public health, specifically obesity, diabetes, and premature mortality. We use US county-level data from the Census of Agriculture, University of Wisconsin Population Health Institute's *County Health Rankings & Roadmaps*, and American Community Survey data from 2012 to examine the local foods and health relationship.

After controlling for numerous community characteristics, we find that there is indeed a statistically significant relationship between access to local foods and health. The result is consistent across several different model specifications. On its face value, this result alone would support the premise that promoting local foods is a strategy to promote better health. But there are two difficulties with that conclusion. First, while statistically significant, the magnitude of the estimated coefficients suggests that the relationships are modest. Second, it is not clear if access to local foods promoted better health or if healthier people create a stronger market for local foods. Indeed, there is growing evidence that building local foods systems with the intent of building more vibrant communities may be short-sighted: building a local foods system will not ensure that people will tap into it, nor that it is equally accessible by all community residents. In the end, we concur with Lev, Hand, and DiGiacomo (2014) that local foods are likely demand driven, and the mere presence of local foods activity is not sufficient to result in healthier community outcomes. This is not to say that local food systems could not be part of a larger set of community strategies, but as with any community economic development strategy, it is not a magic bullet. Additional studies to more rigorously measure associations between local foods and community development, including health outcomes, are needed.

Notes

1. These are focused programs that could be called "pop-up farmers markets" that are organized by local institutions including businesses. For example, as part of a company's wellness program the business will organize a small pop-up farmers market on company property for their employees.
2. USDA Food Atlas: https://www.ers.usda.gov/data-products/food-environment-atlas/ UW Population Health Institute County Health Rankings: https://www.countyhealthrankings.org/.
3. For more details see https://www.countyhealthrankings.org/measure/premature-death-ypll.
4. Principal component analysis, a specific type of factor analysis, is a technique used to identify patterns amongst variables in a dataset and compress them into a single measure. In the simplest sense the structure of a correlation matrix amongst the variables of interest is used to build a weighting scheme to explain the degree to which each variables explains or contribute to the total covariance of all the data. See Jolliffe (2002) for a detailed discussion or Bro and Smilde (2014) for a more introductory discussion.
5. While the obesity and diabetes measures are scaled the same (a percentage), the premature death measure is scaled different. Thus in building the principal component index we standardize the three measures to mean zero with unit variance. When we model the three measures separate, we use the original scaling.
6. https://www.agcensus.usda.gov/Publications/Local_Food/.
7. Within the context of local food systems and social capital see Morales and Suerth (this issue).
8. Notice the local foods index is truncated at around negative two: this is because there are many counties that have zero local foods activity. There are also a handful of "outlier" counties with high concentrations of local foods activity. This characteristic of the local foods index is a main motivation for examining the individual elements of the index (Tables 4).
9. Care must be taken comparing the reported hetrosckedastic corrected standard errors and the reported standardized coefficients. For example, the result on the local food index and premature morality is statistically significant, but the relatively large standard error relative to the standardized coefficient suggests otherwise.
10. For Table 4 we include all of the control variables but suppress reporting them in the table for space consideration. The results on the control variables are largely the same as those reported in Table 3.
11. Some care must be taken with this simple analysis because many local foods are shipped across county boundaries, often flowing from surrounding rural areas into urban markets. Regardless, the relatively modest share of food dollars spent on local foods remains a valid observation.

Disclosure statement

No potential conflict of interest was reported by the authors.

Funding

This work was supported by the Wisconsin Agricultural Experiment State, University of Wisconsin-Madison. The authors are responsible for any errors.

References

Abraido-Lanza, A.F., Chao, M.T., & Florez, K.R. (2005). Do healthy behaviors decline with greater acculturation? Implications for the Latino mortality paradox. *Social Science and Medicine, 61*, 1243–1255. doi:10.1016/j.socscimed.2005.01.016

Adler, N.E., & Ostrove, J.M. (1999). Socioeconomic status and health: What we know and what we don't. *Annals of the New York Academy of Sciences, 896*, 3–15. doi:10.1111/j.1749-6632.1999.tb08101.x

Albrecht, C., Johnson, R., Hamann, S., & Sneyd, L., Ohberg, L., & CoDye, M. (2013). Toward alternative food systems development: Exploring limitations and research opportunities. *Journal of Agriculture, Food Systems, and Community Development, 3*, 151–159. doi:10.5304/jafscd.2013.034.019

Bedimo-Rung, A.L., Mowen, A.J., & Cohen, D.A. (2005). The significance of parks to physical activity and public health: A conceptual model. *American Journal of Preventive Medicine, 28*, 159–168. doi:10.1016/j.amepre.2004.10.024

Berning, J.P. (2012). Access to local agriculture and weight outcomes. *Agricultural and Resource Economics Review, 41*, 57–71. doi:10.1017/S1068280500004184

Bimbo, F., Viscecchia, R., & Nardone, G. (2012). *Does the alternative food supply network affect the human health?* (126th Seminar, June 27–29, 2012, Capri, Italy No. 126060). European Association of Agricultural Economists.

Born, B., & Purcell, M. (2006). Avoiding the local trap scale and food systems in planning research. *Journal of Planning Education and Research, 26*, 195–207. doi:10.1177/0739456X06291389

Bro, R., & Smilde, A.K. (2014). Principal component analysis. *Anal. Methods, 6*, 2812–2831. doi:10.1039/C3AY41907 J

Brunekreef, B., & Holgate, S.T. (2002). Air pollution and health. *The Lancet, 360*, 1233–1242. doi:10.1016/S0140-6736(02)11274-8

Brunello, G., Fort, M., Schneeweis, N., & Winter-Ebmer, R. (2016). The causal effect of education on health: What is the role of health behaviors? *Health Economics, 25*, 314–336. doi:10.1002/hec.3141

Casey, M.M., Call, K.T., & Klingner, J.M. (2001). Are rural residents less likely to obtain recommended preventive healthcare services? *American Journal of Preventive Medicine, 21*, 182–188. doi:10.1016/S0749-3797(01)00349-X

Caspi, C.E., Sorensen, G., Subramanian, S.V., & Kawachi, I. (2012). The local food environment and diet: A systematic review. *Health & Place, 18*, 1172–1187. doi:10.1016/j.healthplace.2012.05.006

Cattell, V. (2001). Poor people, poor places, and poor health: The mediating role of social networks and social capital. *Social Science and Medicine, 52*, 1501–1516. doi:10.1016/S0277-9536(00)00259-8

Centers for Disease Control and Prevention. (2011). *Strategies to prevent obesity and other chronic diseases: The CDC guide to strategies to increase the consumption of fruits and vegetables.* Atlanta, GA: US Department of Health and Human Services.

Centers for Disease Control and Prevention. (2014). *National diabetes statistics report: Estimates of diabetes and its burden in the United States, 2014.* Atlanta, GA: US Department of Health and Human Services.

Churchill, S.A., Ocloo, J.E., & Siawor-Robertson, D. (2016). Ethnic diversity and health outcomes. *Social Indicators Research*. Early access version. 1–36. doi:10.1007/s11205-016-1454-7

Clancy, K. (2013). Bringing a systems approach to food systems: High-priority research approaches for transforming US food systems. *Journal of Agriculture, Food Systems and Community Development, 3*, 5–7. doi:10.5304/jafscd.2013.034.021

Coleman-Jensen, A., Rabbitt, M., Gregory, C., & Singh, A. (2016). *Household food security in the United States in 2015* (ERR-215). Washington, DC: US Department of Agriculture, Economic Research Service. doi:10.2139/ssrn.2504067

Community Science. (2013). *SNAP healthy food incentives cluster evaluation: 2013 final report*. Retrieved from https://ecologycenter.org/wp-content/uploads/2013/11/2013-Cluster-Evaluation-Final-Report_-final_10.4.13.pdf

Creamer, N.G., & Dunning, R.D. (2012). Local food systems for a healthy population. *North Carolina Medical Journal, 73*, 310–314.

Crilly, J.F., Keefe, R.H., & Volpe, F. (2011). Use of electronic technologies to promote community and personal health for individuals unconnected to health care systems. *American Journal of Public Health, 101*, 1163–1167. doi:10.2105/AJPH.2010.300003

Darby, K., Batte, M.T., Ernst, S., & Roe, B. (2008). Decomposing local: A conjoint analysis of locally produced foods. *American Journal of Agricultural Economics, 90*, 476–486. doi:10.1111/j.1467-8276.2007.01111.x

Deller, S., Canto, A., & Brown, L. (2015). Rural poverty, health and food access. *Regional Science Policy & Practice, 7*, 61–74. doi:10.1111/rsp3.12056

Deller, S.C., & Deller, M.A. (2010). Rural crime and social capital. *Growth and Change., 41*, 221–275. doi:10.1111/j.1468-2257.2010.00526.x

Derkatch, C., & Spoel, P. (2016). Public health promotion of "local food": Constituting the self-governing citizen-consumer. *Health, 3*, 46–47. doi:10.15353/cfs-rcea.v3i1.144

Dunn, B. (2013). *Increasing access to local food: Policies from other places as a guide to increasing local food access through land use planning in Ontario*. Policies from the Field: Promising Food Polices from Other Places. Toronto: Sustain Ontario.

Dunne, J.B., Chambers, K.J., Giombolini, K.J., & Schlegel, S.A. (2010). What does 'local' mean in the grocery store? Multiplicity in food retailers' perspectives on sourcing and marketing local foods. *Renewable Agriculture and Food Systems, 26*, 46–59. doi:10.1017/S1742170510000402

Durham, C.A., King, R.P., & Roheim, C.A. (2009). Consumer definitions of 'locally grown' for fresh fruits and vegetables. *Journal of Food Distribution Research, 40*, 56–62.

Emery, M., & Flora, C. (2006). Spiraling-up: Mapping community transformation with community capitals framework. *Community Development, 37*, 19–35. doi:10.1080/15575330609490152

Feenstra, G.W. (1997). Local food systems and sustainable communities. *American Journal of Alternative Agriculture, 12*, 28–36. doi:10.1017/S0889189300007165

Feenstra, G.W. (2002). Creating space for sustainable food systems: Lessons from the field. *Agriculture and Human Values, 19*, 99–106. doi:10.1023/A:1016095421310

Ferrer, M.C., Fonsah, E.G., Ramirez, O., & Escalante, C.L. (2011, July 24–26). *Local food impact on health and nutrition*. Paper presented at the Agricultural and Applied Economics Association Annual Meetings, Pittsburg, PA.

Flegal, K.M., Kruszon-Moran, D., Carroll, M.D., Fryar, C.D., & Ogden, C.L. (2016). Trends in obesity among adults in the United States, 2005 to 2014. *Journal of the American Medical Association, 315*, 2284–2291. doi:10.1001/jama.2016.6458

Freedman, D.A., Mattison-Faye, A., Alia, K., Guest, M.A., & Hébert, J.R. (2014). Comparing farmers' market revenue trends before and after the implementation of a monetary incentive for recipients of food assistance. *Preventing Chronic Disease, 11*, 130347. doi:10.5888/pcd11.130347

Fung, T.T., Chiuve, S.E., McCullough, M.L., Rexrode, K.M., Logroscino, G., & Hu, F.B. (2008). Adherence to a DASH-style diet and risk of coronary heart disease and stroke in women. *Archives of Internal Medicine, 168*, 713. doi:10.1001/archinte.168.7.713

Garnett, T. (2013). Food sustainability: Problems, perspectives and solutions. *Proceedings of the Nutrition Society, 72*, 29–39. doi:10.1017/S0029665112002947

Glanz, K., Sallis, J.F., Saelens, B.E., & Frank, L.D. (2005). Healthy nutrition environments: Concepts and measures. *American Journal of Health Promotion, 19*, 330–333. doi:10.4278/0890-1171-19.5.330

Goodman, M. (2013). Wrapping and stuffing food relationally: Pleasure, place, production and power. In M. McWilliams (Ed.), *Wrapped and stuffed foods: Proceeedings of the oxford symposium on food and cookery 2012* (pp. 52–56). Devon: Prospect Books.

Guptill, A., & Wilkins, J. (2002). Buying into the food system: Trends in food retailing in the US and implications for local foods. *Agriculture and Human Values, 19*, 39–51. doi:10.1023/A:1015024827047

Hamm, M. (2008). Linking sustainable agriculture and public health: Opportunities for realizing multiple goals. *Journal of Hunger and Environmental Nutrition, 3*, 169–185. doi:10.1080/19320240802243241

Hardesty, S.D. (2008). The growing role of local food markets. *American Journal of Agricultural Economics, 90*, 1289–1295. doi:10.1111/j.1467-8276.2008.01219.x

He, F.J., Nowson, C.A., & MacGregor, G.A. (2006). Fruit and vegetable consumption and stroke: Meta-analysis of cohort studies. *The Lancet, 367*, 320–326. doi:10.1016/S0140-6736(06)68069-0

Herman, D.R., Harrison, G.G., Afifi, A.A., & Jenks, E. (2008). Effect of a targeted subsidy on intake of fruits and vegetables among low-income women in the special supplemental nutrition program for women, infants, and children. *American Journal of Public Health, 98*, 98–105. doi:10.2105/AJPH.2005.079418

Heron, M. (2016). *Deaths: Leading causes for 2014* (National Vital Statistics Reports). Hyattsville, MD: National Center for Health Statistics, 65.

Hu, F.B. (2003). Plant-based foods and prevention of cardiovascular disease: An overview. *The American Journal of Clinical Nutrition, 78*, 544S–551S.

Jablonski, B.B.R., McFadden, D., & Colpaart, A. (2016). Analyzing the role of community and individual factors in food insecurity: Identifying diverse barriers across clustered community members. *Journal of Community Health, 41*, 910–923. doi:10.1007/s10900-016-0171-0

Jilcott, S.B., Wade, S., McGuirt, J.T., Wu, Q., Lazorick, S., & Moore, B.J. (2011). The association between the food environment and weight status among eastern North Carolina youth. *Public Health and Nutrition, 14*, 1610–1617. doi:10.1017/S1368980011000668

Jolliffe, I. (2002). *Principal component analysis*. Hoboken, NJ: Wiley.

Joshi, A., Azuma, A.M., & Feenstra, G. (2008). Do farm-to-school programs make a difference? Findings and future research needs. *Journal of Hunger & Environmental Nutrition, 3*, 229–246. doi:10.1080/19320240802244025

Kawachi, I., Subramanian, S.V., & Kim, D.L. (2007). Social capital and health. In I. Kawachi, S.V. Subramanian, & D. Kim (Eds.), *Social capital and health* (pp. 1–28). New York, NY: Springer.

Keene, A., & Deller, S.C. (2015). Evidence of the environmental Kuznets' curve among US counties and the impact of social capital. *International Regional Science Review, 38*, 358–387. doi:10.1177/0160017613496633

Kelly, B., Flood, V.M., & Yeatman, H. (2011). Measuring local food environments: An overview of available methods and measures. *Health & Place, 17*, 1284–1293. doi:10.1016/j.healthplace.2011.08.014

King, R.P., Hand, M.S., DiGiacomo, G., Clancy, K., Gomez, M.I., Hardesty, S.D., … McLaughlin, E.W. (2010). *Comparing the structure, size and performance of local and mainstream food supply chains* (USDA ERS Report Number 99). Retrieved from https://www.ers.usda.gov/webdocs/publications/46405/7028_err99_reportsummary_1_.pdf?v=41056

Kropf, M.L., Holben, D.H., Holcomb, J.P., & Anderson, H. (2007). Food security status and produce intake and behaviors of special supplemental nutrition program for women, infants, and children and farmers' market nutrition program participants. *Journal of the American Dietetic Association, 107*, 1903–1908. doi:10.1016/j.jada.2007.08.014

Lamie, R.D., Dunning, R., Bendfeldt, E., Massey Lelekacs, J., Velandia, M., & Meyer, L. (2013). Local food systems in the South: A call for a collaborative approach to assessment. *Choices, 28*, 1–5.

Lev, L., Hand, M.S., & DiGiacomo, D. (2014). What does local deliver? In R.P. King, M.S. Hand, & M.I. Gómez (Eds.), *Growing local: Case studies on local food supply chains* (pp. 291–312). Lincoln: University of Nebraska Press.

Lindsay, S., Lambert, J., Penn, T., Hedges, S., Ortwine, K., Mei, A., … Wooten, W.J. (2013). Monetary matched incentives to encourage the purchase of fresh fruits and vegetables at farmers' markets in underserved communities. *Preventing Chronic Disease, 10*, 13124. doi:10.5888/pcd10.130124

Low, S.A., & Vogel, S. (2011). *Direct and intermediated marketing of local foods in the United States* (Research Report Number 128). Washington, DC: US Department of Agriculture, Economic Research Service. doi: 10.2139/ssrn.2114361

Mackenbach, J.D., Lakerveld, J., van Lenthe, F.J., Kawachi, I., McKee, M., Rutter, H., … Oppert, J.M. (2016). Neighbourhood social capital: Measurement issues and associations with health outcomes. *Obesity Reviews, 17*, 96–107. doi:10.1111/obr.12373

Mansfield, C., & Novick, L.F. (2012). Poverty and health: Focus on North Carolina. *North Carolina Medical Journal, 73*, 366–373.

Marsden, T., Banks, J., & Bristow, G. (2000). Food supply chain approaches: Exploring their role in rural development. *Sociologia Ruralis, 40*, 424–438. doi:10.1111/1467-9523.00158

Martinez, S., Hand, M., Da Pra, M., Pollack, S., Ralston, K., Smith, T., … Newman, C. (2010). *Local food systems: Concepts, impacts, and issues* (ERS Report Number 97). Retrieved from https://www.ers.usda.gov/webdocs/publications/46393/7053_err97_reportsummary_1_.pdf?v=42265

McCormack, L.A., Laska, M.N., Larson, N.I., & Story, M. (2010). Review of the nutritional implications of farmers' markets and community gardens: A call for evaluation and research efforts. *Journal of the American Dietetic Association, 110*, 399–408. doi:10.1016/j.jada.2009.11.023

McFadden, D.T. (2015). What do we mean by "local foods?" *Choices, 30*, 1–6.

Minaker, L.M., Raine, K.D., Fisher, P., Thompson, M.E., Van Loon, J., & Frank, L.D. (2014). Food purchasing from farmers' markets and community-supported agriculture is associated with reduced weight and better diets in a population-based sample. *Journal of Hunger & Environmental Nutrition, 9*, 485–497. doi:10.1080/19320248.2014.898175

Montgomery, L.E., Kiely, J.L., & Pappas, G. (1996). The effects of poverty, race, and family structure on US children's health: Data from the NHIS, 1978 through 1980 and 1989 through 1991. *American Journal of Public Health, 86*, 1401–1405. doi:10.2105/AJPH.86.10.1401

Montonen, J., Knekt, P., Järvinen, R., & Reunanen, A. (2004). Dietary antioxidant intake and risk of type 2 diabetes. *Diabetes Care, 27*, 362–366.

Morland, K.B., & Evenson, K.R. (2009). Obesity prevalence and the local food environment. *Health & Place, 15*, 491–495. doi:10.1016/j.healthplace.2008.09.004

Murray, S. (2006). Poverty and health. *Canadian Medical Association Journal, 174*(7), 923.

Ogden, C.L., Carroll, M.D., Lawman, H.G., Fryar, C.D., Kruszon-Moran, D., Kit, B.K., & Flegal, K.M. (2016). Trends in obesity prevalence among children and adolescents in the United States, 1988–1994 through 2013–2014. *Journal of the American Medical Association, 315*, 2292–2299. doi:10.1001/jama.2016.6361

Pappas, G., Queen, S., Hadden, W., & Fisher, G. (1993). The increasing disparity in mortality between socioeconomic groups in the United States, 1960 and 1986. *New England Journal of Medicine, 329*, 103–109. doi:10.1056/NEJM199307083290207

Penchansky, R., & Thomas, J.W. (1981). The concept of access: Definition and relationship to consumer satisfaction. *Medical Care, 19*, 127–140. Retrieved from https://www.jstor.org/stable/3764310

Pickett, K.E., & Wilkinson, R.G. (2015). Income inequality and health: A causal review. *Social Science & Medicine, 128*, 316–326. doi:10.1016/j.socscimed.2014.12.031

Racine, E.F., Smith Vaughn, A., & Laditka, B.S. (2010). Farmers' market use among African-American women participating in the special supplemental nutrition program for women, infants, and children. *Journal of the American Dietetic Association, 110*, 441–446. doi:10.1016/j.jada.2009.11.019

Rolls, B.J., Ello-Martin, J.A., & Tohill, B.C. (2004). What can intervention studies tell us about the relationship between fruit and vegetable consumption and weight management? *Nutrition Reviews, 62*, 1–17. doi:10.1111/j.1753-4887.2004.tb00001.x

Ruge, D., & Mikkelsen, B.E. (2013). Local public food strategies as a social innovation: Early insights from the LOMA-Nymarkskolen case study. *Acta Agriculturae Scandinavica, Section B – Soil & Plant Science, 63*, 56–65. doi:10.1080/09064710.2013.793736

Rupasingha, A., Goetz, S.J., & Freshwater, D. (2006). The production of social capital in US counties. *The Journal of Socio-economics, 35*, 83–101. doi:10.1016/j.socec2005.11.001

Salois, M.J. (2012). Obesity and diabetes, the built environment, and the 'local' food economy in the United States, 2007. *Economics & Human Biology, 10*, 35–42. doi:10.1016/j.ehb.2011.04.001

Segal, L.M., Rayburn, J., & Martín, A. (2016). *The state of obesity: Better policies for a healthier America, 2016*. Retrieved from https://healthyamericans.org/assets/files/TFAH-2016-ObesityReport-FINAL.pdf

Seguin, R.A., Morgan, E.H., Hanson, K.L., Ammerman, A.S., Jilcott Pitts, S.B., Kolodinsky, J., ... Sitaker, Marilyn. (2017). Farm fresh foods for healthy kids (F3HK): An innovative community supported agriculture intervention to prevent childhood obesity in low-income families and strengthen local agricultural economies. *BMC Public Health, 17*, 306. doi:10.1186/s12889-017-4202-2

Shaffer, R., Deller, S.C., & Marcouiller, D.W. (2006). Rethinking community economic development. *Economic Development Quarterly, 20*, 59–74. doi:10.1177/0891242405283106

Stewart, M.A. (1995). Effective physician-patient communication and health outcomes: A review. *CMAJ: Canadian Medical Association Journal, 152*, 1423–1433.

Story, M., Kaphingst, K.M., Robinson-O'Brien, R., & Glanz, K. (2008). Creating healthy food and eating environments: Policy and environmental approaches. *Annual Review of Public Health, 29*, 253–272. doi:10.1146/annurev.publhealth.29.020907.090926

Sundbo, D.I.C. (2013). Local food: The social construction of a concept. *Acta Agriculturae Scandinavica, Section B – Soil & Plant Science, 63*, 66–77. doi:10.1080/09064710.2013.794857

Szreter, S., & Woolcock, M. (2004). Health by association? Social capital, social theory, and the political economy of public health. *International Journal of Epidemiology, 33*, 650–667. doi:10.1093/ije/dyh013

Taylor, J.C., & Johnson, R.K. (2013). Farm to school as a strategy to increase children's fruit and vegetable consumption in the United States: Research and recommendations: Fruit and vegetable consumption in farm to school. *Nutrition Bulletin, 38*, 70–79. doi:10.1111/nbu.12009

Tohill, B.C., Seymour, J., Serdula, M., Kettel-Khan, L., & Rolls, B.J. (2004). What epidemiologic studies tell us about the relationship between fruit and vegetable consumption and body weight. *Nutrition Reviews, 62*, 365–374. doi:10.1111/j.1753-4887.2004.tb00007.x

Truesdale, B.C., & Jencks, C. (2016). Income inequality and health: Strong theories, weaker evidence. *Frontiers in Public Health Services and Systems Research, 5*, 30–37. doi:10.13023/FPHSSR.0505.05

University of Wisconsin Population Health Institute. (2013). *County health rankings and roadmaps*. Madison, WI: University of Wisconsin. Retrieved from https://www.countyhealthrankings.org/

Ver Ploeg, M. (Ed.). (2012). *Access to affordable and nutritious food: Measuring and understanding food deserts and their consequences: Report to Congress*. Washington, DC: US Department of Agriculture, Economic Research Service.

Vila, P.M., Booske, B.C., & Remington, P.L. (2006). *Measuring mortality in the Wisconsin county health rankings*. Madison, WI: University of Wisconsin Population Health Institute.

Walker, R.E., Keane, C.R., & Burke, G.J. (2010). Disparities and access to healthy food in the United States: A review of food deserts literature. *Health & Place, 16*, 876–884. doi:10.1016/j.healthplace.2010.04.013

Webber, K.H., Stephenson, T.J., Mayes, L., & Stephenson, L. (2013). Nutrition knowledge and dietary habits of farmers market patrons. *World Applied Sciences Journal, 23*, 267–271. doi:10.5829/idosi.wasj.2013.23.02.1652

Wholesome Wave. (2012). *Double value coupon program diet and shopping behavior study*. Retrieved from https://www.oecd.org/site/agrfcn/Double-Value-Coupon-Program-Diet-Shopping-Behavior-Study-.pdf

Winkleby, M.A., Jatulis, D.E., Frank, E., & Fortmann, S.P. (1992). Socioeconomic status and health: How education, income, and occupation contribute to risk factors for cardiovascular disease. *American Journal of Public Health, 82*, 816–820. doi:10.2105/AJPH.82.6.816

Wulff, K., Donato, D., & Lurie, N. (2015). What is health resilience and how can we build it? *Annual Review of Public Health, 36*, 361–374. doi:10.1146/annurev-publhealth-031914-122829

Young, C.R., Aquilante, J.L., Solomon, S., Colby, L., Kawinzi, M.A., Uy, N., & Mallya, G. (2013). Improving fruit and vegetable consumption among low-income customers at farmers' markets: Philly food bucks, Philadelphia, Pennsylvania, 2011. *Preventing Chronic Disease, 10*, 1–8. doi:10.5888/pcd10.120356

Zepeda, L., & Li, J. (2006). Who buys local food? *Journal of Food Distribution Research, 37*, 1–11.

Zepeda, L., & Nie, C. (2012). What are the odds of being an organic or local food shopper? Multivariate analysis of US food shopper lifestyle segments. *Agriculture and Human Values, 29*, 467–480. doi:10.1007/s10460-012-9364-z

Ziersch, A.M., Baum, F.E., MacDougall, C., & Putland, C. (2005). Neighbourhood life and social capital: The implications for health. *Social Science & Medicine, 60*, 71–86. doi:10.1016/j.socscimed.2004.04.027

The impacts of the Farmers Market and Local Food Promotion programs

Jeffrey K. O'Hara and Carlos Coleman

ABSTRACT

Farmers Market Promotion Program (FMPP) and Local Food Promotion Program grants are awarded by the US Department of Agriculture to bolster local food markets. We assess the impacts of these programs by analyzing responses to a national survey of farmers market administrators, undertaking two case studies of grant recipients, and reviewing previous literature. We find that there is a high association of FMPP recipient markets with greater organizational administrative capacity and improved market performance between 2013 and 2014. Our case studies reveal how grants can be effective when the projects have strong stakeholder support and collaborate with other community organizations. We also describe how some of the pilot projects supported by FMPP led to more targeted local food grant programs.

Introduction

Farmers undertake "local" food sales when they sell products at direct-to-consumer (DTC) agricultural markets, which can include farmers markets, on-farm or roadside stands, and community-supported agriculture programs. They also occur when farmers sell directly to retail institutions and intermediaries such as restaurants, grocery stores, hospitals, schools, universities, and distributors that market locally branded products. In 2015, the value of local food sales from farmers was $8.7 billion (US Department of Agriculture National Agricultural Statistics Service [USDA NASS], 2016). Local food markets can provide market opportunities for farmers in either urban or rural locations. Of the 167,009 farmers that were estimated to be selling in local food markets in 2015, 53% were located in metropolitan counties (USDA NASS, 2016).

Both the economic development potential of local food markets and impediments to their success have motivated the use of public support for them. The objective of this article is to describe two federal grant programs, the Farmers Market Promotion Program (FMPP) and Local Food Promotion Program (LFPP), and to assess what can be concluded about their impacts given the available evidence. FMPP and LFPP are explicitly intended to develop, improve, and expand local food markets, as well as to support outreach, training, and technical assistance for such purposes (US Senate Committee on Agriculture, Nutrition, & Forestry

[US Senate], 2014). Both programs are administered by the US Department of Agriculture's Agricultural Marketing Service (USDA AMS). FMPP grants provide support for DTC agricultural marketing outlets, while LFPP grants support food business enterprises that are involved with processing, aggregating, or distributing local food products for institutional purchases.

Despite their strategic importance, there are several impediments with evaluating the impacts of these two grant programs. One issue is that FMPP and LFPP awards are administered without randomized control and treatment groups. This implies that grant recipients may have systematically different levels of organizational capacity than non-recipient organizations, the latter of which may not have awareness of the programs and/or the resources to submit a successful application. Also, the awards can support disparate kinds of project activities, with many awards supporting more than one type of project activity, and the intended outcomes of each respective activity might be evaluated using different metrics. Similarly, grants are implemented on a non-standard spatial scale. For instance, one FMPP award may support an individual farmers market, whereas another FMPP award could support statewide training. Further, there are other programs at federal, state, and local levels that can be used to enhance local food markets that may be interacting with FMPP and LFPP awards (Hughes & Boys, 2015). Yet another evaluation challenge is that some of the longer term impacts from implementing the grant may not occur until after the relatively short project and reporting period. While the focus of our article specifically focuses on local food subsidies, the same kinds of evaluation challenges also exist with other multifaceted/non-standard economic development subsidies that are non-randomly awarded.

Nonetheless, our review of these grant programs is essential given the longevity of FMPP, the specific objective of FMPP and LFPP at supporting local food markets (as opposed to other programs that could support local food programs but do not have an explicit mandate to do so), and because these programs will require reauthorization to continue upon expiration of the 2014 Farm Bill. Further, there is importance in disseminating the lessons learned from the experience of awardees to-date so that other prospective applicants can ascertain what kinds of outcomes could be accomplished with the grant's support, and learn about solutions that applicants developed when confronted with unanticipated challenges during the implementation stages of the grant administration. Another objective in discussing some of the details of particular projects is to support the development of a community of practice in measuring the performance of local food systems. This could ultimately contribute to the creation of appropriate standardized reporting metrics for grant recipients so that they can be aggregated to disseminate the broader grant impacts.

Planners in communities and regions contemplating investing in local food systems as an economic development strategy must compare the merits of supporting local food systems relative to other opportunities. Local food systems have appeal as an economic development strategy since they can attain several of the five objectives that such strategies should ideally accomplish (Hustedde, Shaffer, & Pulver, 2005):

- Attracting new, basic, or export employers.
- Increasing the share of local products sold in preexisting markets.
- Encouraging the start-up of new firms.
- Helping existing firms become more efficient.
- Using aids and programs from broader levels of government.

A well-designed local food grant application has the potential to accomplish all of these objectives. It uses the fifth objective – obtain higher levels of government support – to capture an existing market (food) that can benefit new or existing employers and firms. Specific reasons why local food markets are perceived as compelling economic development strategies are that first, DTC agricultural sales occur primarily from small- and medium-sized farms (Ahearn & Sterns, 2013; Low & Vogel, 2011). These farms may otherwise be unable or unwilling to sell products through conventional agricultural supply chains. Further, DTC agricultural markets can enable farmers and vendors to improve entrepreneurial and business management skills (Feenstra, Lewis, Hinrichs, Gillespie, & Hilchey, 2003). Second, local food farms may be relatively more labor-intensive, enhancing local employment activities and opportunities. For example, fruit and vegetable farms that sell locally employ 13 employees per million dollar of sales, whereas non-local fruit and vegetable farms employ three workers for this sales level (Low & Vogel, 2011). Third, the localized economic impacts of food purchased directly from farmers can be greater than traditional food purchases at grocery stores (Hughes, Brown, Miller, & McConnell, 2008; Hughes & Isengildina-Massa, 2015). This occurs because the food products at farmers markets are more likely to be local relative to grocery store products, and also because farmers that sell food through local markets may hire labor and purchase inputs locally to a greater extent than other farmers (Jablonski & Schmit, 2014).

Despite the benefits that farmers markets and other DTC agricultural venues can provide to communities, economic and resource constraints can contribute to management challenges. In particular, Stephenson, Lev, and Brewer (2008) found that a small size, need for vendors, high level of manager turnover, low budget, and dependence on volunteer labor were factors associated with the failure of a farmers market. Ragland and Tropp (2009) found in 2006 that 59% of farmers markets rely exclusively on volunteer labor, that 39% had a paid manager with no other employees, and that farmers markets had a mean operating budget of $20,864. Schmit and Gomez (2011) found that public support for farmers markets can be particularly critical in rural communities. In particular, the need for public assistance for DTC agricultural markets to serve low-income communities has arisen since the conversion of the paper-based food stamp program to the electronic Supplemental Nutrition Assistance Program (SNAP) program presented administrative challenges in redeeming food assistance benefits at farmers markets and from direct marketing farmers.

Thus, given both the program evaluation challenges with FMPP and LFPP and the importance of being able to assess what we conclude about their impacts, we piece together several disparate sources of information. Specifically, we use data from a farmers market manager survey to compare the attributes and performance of markets that had received an FMPP grant with those that had not, undertake a review of previous literature, and report aggregated LFPP performance report evaluation metrics. We further provide case studies of one FMPP and one LFPP recipient. We find that there is a high association of FMPP recipient markets with greater organizational administrative capacity and improved market performance between 2013 and 2014. However, we caution against a causal interpretation of these correlations due to the aforementioned issue of the grants being non-randomly awarded. We also find in our case studies that critical aspects for the success of implementing a grant are ensuring that there is stakeholder support for the proposal's objectives, as well as collaborations with other organizations in the community. We also describe how some of the pilot projects supported by FMPP led to the development of more targeted local food grant programs and that the future effectiveness of FMPP and LFPP may depend on whether the increased organizational capability that some local food organizations experienced in the past decade can enable them to successfully administer larger grants.

The article provides a detailed overview of FMPP and LFPP, describes grant-writing resources to prospective applicants, discusses other local food grant programs, and identifies planning templates and guidance available to communities interested in knowing whether applying for a local food grant is worthwhile. The subsequent sections present the data analysis, feature case studies, and offer concluding comments.

Background

FMPP and LFPP overview

Although the FMPP program name contains the phrase "farmers market," FMPP awards can be used to improve, expand, and develop any kind of DTC marketing channel, including farmers markets, roadside stands, community-supported agriculture (CSA) programs, and on-farm pick-your-own operations. Applicants eligible for FMPP funds include a wide variety of community-based organizations. Such organizations could be agricultural cooperatives, agricultural businesses, local governments, nonprofits, producer networks or associations, public benefit corporations, economic development corporations, regional farmers market associations, and Tribal governments.

LFPP was established to support food business enterprises that process, distribute, aggregate, or store local or regional food products. LFPP awards can support projects that either move a product from a producer to a distributor/food hub, or move a product from a distributor/food hub to a retail outlet. In this context, a "food hub" is an organization that manages the aggregation, distribution, and marketing of source-identified food products that primarily come from local and regional producers. FMPP and LFPP were directed to prioritize applicants located in areas of concentrated poverty with limited access to fresh local foods (as defined by the USDA Economic Research Service Food Access Research Atlas). There are two types of LFPP grants: "implementation" grants to establish a local or regional food business enterprise and "planning" grants for market research, business planning, and feasibility studies. Further detail about the two programs is contained in Table 1.

USDA AMS disbursed 958 FMPP awards that totaled \$73 million (in nominal dollars) between 2006 and 2016. FMPP projects have been awarded in all 50 states and in the District of Columbia. In Figure 1, we see that USDA AMS awarded 20 FMPP grants each in 2006 and 2007 for a total of \$2 million. From 2008 through 2012, USDA AMS awarded 522 FMPPs totaling \$30 million. No FMPP grants were awarded during 2013. Between 2014 and 2016, USDA AMS awarded \$41 million in FMPP grants and an additional \$38 million in LFPP grants. The decline in 2016 of the number of FMPP and LFPP awards without a corresponding decline in the value of awards reflects an increase in the maximum award size between 2015 and 2016.

FMPP grants have been awarded to single markets, regional networks of markets, and organizations operating at a statewide or multistate level. A single FMPP award can support multiple types of project activities. The most common include promotional campaigns that educate and inform the public about DTC agricultural markets, offer technical assistance and training to DTC agricultural farmers, and assist farmers markets with serving low-income

Table 1. FMPP and LFPP background information.

	Farmers Market Promotion Program	Local Food Promotion Program
Farm bill authorizations	2002, 2008, 2014	2014
First-year program administered	2006	2014
Objective	Improve, expand, or develop DTC agricultural markets (farmers markets, roadside or on-farm stands, CSAs)	Support food businesses that process, distribute, aggregate, or store local food products
Match requirements	None	25%
Grant caps	"Capacity building": $250,000; "Community development, training, and technical assistance": $500,000	"Planning": $100,000; "Implementation": $500,000

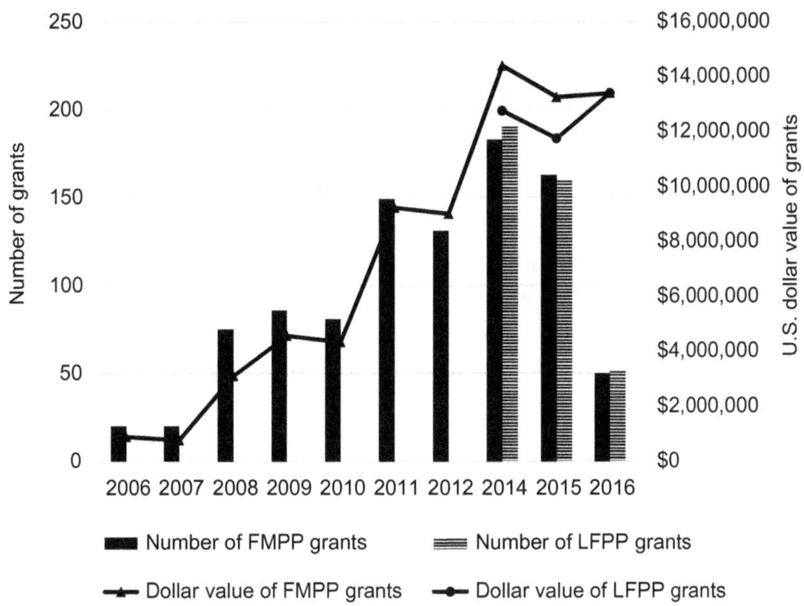

Figure 1. Number and US dollar value of FMPP and LFPP grants by year.
Note: Figure created by authors with data from USDA AMS (2016a).

communities. Other project activities that FMPP awards support include establishing new markets, procuring supplies and equipment (such as a freezer), undertaking market research to assess the feasibility of DTC agricultural marketing, improving transportation to markets for vendors and consumers, developing agritourism projects, and establishing new organizational partnerships.[1]

USDA AMS developed technical assistance for prospective grant writers in 2015 in order to make FMPP and LFPP grants more accessible, reduce application costs, and improve the award performance of recipients.[2] Applicants must submit a project narrative, a budget, organizational qualifications, and support letters from partnering organizations. Applicants are further required to have a "DUNS Number" (which is a unique identification number for their business), register with "SAM" as a federal contractor, and register with grants.gov (the electronic interface for submitting a grant proposal).

Evaluations of local food grant programs

One source of information about FMPP's impacts is USDA AMS publically posted individual performance reports from grant recipients (USDA AMS, 2016b). A more comprehensive assessment of FMPP's impacts was undertaken by Miller and Roper (2013), who surveyed FMPP recipients that were awarded grants between 2006 and 2011. They received 200 responses from grantees. They found that, on average, sales increased by 27% at markets that had received FMPP awards, the number of participating producers increased by 34%, customer counts increased by 47%, and the number of first-time customers increased in 94% of markets. Miller and Roper (2013) further found that using FMPP funds to enhance professional skills of farmers market managers and staff was correlated with an increase of vendor sales at the market of $85,366.

Between 2006 and 2012, FMPP performance reports did not have a standard template. They predominately consisted of project narratives, with considerable variation in the detail provided by applicants. USDA AMS developed standard reporting forms for both FMPP and LFPP in 2014 for the purposes of being able to aggregate the impacts of the projects. USDA AMS compiled the responses of organizations that completed 2014 LFPP performance reports. These projects consisted of $1 million awarded to 45 LFPP "planning" grants and $4.3 million awarded to 49 LFPP "implementation" grants. The recipients of LFPP implementation grants reported that their projects increased sales by $14.9 million, created 362 direct jobs, created 469 indirect jobs, retained 730 jobs, and benefited 2386 farmers/producers. The recipients further reported that the awards expanded 583 markets and established 640 markets.

Other studies have examined the effectiveness of statewide local food programs. For example, Hughes and Isengildina-Massa (2015) evaluated the impacts of a "buy local" policy in South Carolina. They found that the policy resulted in an 11% increase in farmers market sales and 9% increase in profitability in 2011, resulting in a net increase of $751,000 increase in output, $104,000 increase in income, and 26 jobs. They also concluded that value-added processing could be a critical way to increase economic impacts from buy local policies.

Other local food subsidies

FMPP supported a diverse portfolio of project activities in its early years, as it was one of the first federal grant programs specifically intended for local food markets. Thus, some of the successes that were demonstrated through pilot FMPP projects led to the development of programs with more targeted and specific support for local food markets. Three such examples include:

- The 2008 Farm Bill specified FMPP to devote at least 10% of funds to support the installation of electronic benefit transfer (EBT) machines that accept SNAP benefits at farmers markets. However, USDA's Food and Nutrition Service (FNS) subsequently made additional funding, equipment, and resources available to support EBT machines at local food markets. As a consequence of these cumulative efforts, 3500 farmers markets and 3496 direct marketing farmers were authorized SNAP retailers as of 2016, and they accepted $20.2 million in SNAP benefits (US Department of Agriculture Food & Nutrition Service [USDA FNS], 2017). In part due to FNS's efforts, FMPP no longer had an explicit mandate to support EBT installation when it was reauthorized in the 2014

Farm Bill. Further, AMS and FNS collaborate to ensure that their respective efforts are supporting distinct project activities.

- Offering financial incentives for SNAP redemptions was a complementary innovation that was pioneered at farmers markets within the past decade. Some of the market organizations that managed these programs had received administrative support from FMPP awards. These farmers market incentive programs served as a model for the federal Food Insecurity Nutrition Incentive program, which was established in 2014 to subsidize the redemption of SNAP benefits for fruits and vegetables.
- FMPP awards occasionally provided support for the institutional purchases of local food by retail institutions (schools, hospitals, universities, restaurants, and grocery stores) in FMPP's earlier years. Federal grant programs specifically intended for such markets were subsequently developed. For example, a competitive Farm to School Program was established within FNS to provide grants for program planning, implementation, and activities that support farm-to-school programs. Between 2013 and 2015, FNS awarded $15.1 million to 221 farm-to-school applicants (Benson, Russell, & Kane, 2015). Similarly, LFPP was subsequently developed to support other kinds of non-direct local food market investments.

Other USDA grant and loan programs can be used to support local and regional food systems. However, unlike FMPP and LFPP, many federal grant and loan programs are not specifically designed to enhance local and regional food systems per se. This makes it challenging to quantify and aggregate the value of federal subsidies supporting local and regional food systems.

State governments have also provided resources for local food markets either through administering federal grants (e.g. Specialty Crop Block Grants), providing match funds to augment federal grants (e.g. Federal State Marketing Improvement Program), or providing financial support directly. For example, "buy local" state agricultural marketing programs, which as of 2010 existed in each state, have been funded predominately by state legislatures and/or with federal support (Onken & Bernard, 2010). Municipalities, foundations, and the health care sector have also provided considerable financial assistance for the development of local and regional food markets.

Economic development planning resources for local foods

In some communities it may not be readily apparent whether applying for local food grants like FMPP or LFPP is a worthwhile use of resources, particularly if it is unclear whether there is a sufficient level of community engagement to implement local food projects. There are resources available to communities that provide structure and guidance to evaluate whether local foods are a promising economic development opportunity. We highlight three such examples of economic development planning resources that can inform the viability of applying for an FMPP or LFPP.

The Community Economic Development Strategy (CEDS) is one such example of a regional economic development plan. A CEDS identifies issues relating to economic development challenges, opportunities, objectives, and stakeholder engagement in a region. A CEDS is required to apply for assistance from the Economic Development Administration. Another such example of a structured economic development planning process is Stronger Economies

Together (SET). SET is a process for multi-county regions in rural areas to collaborate and develop economic plans that enhance their region's comparative advantages. SET is administered by the Southern Rural Development Center, which provides coaching and technical assistance to regional teams during the development and implementation of the plan. Several SET initiatives have featured local food systems as a central component of their initiative (Stronger Economies Together [SET], 2016).

A third recently developed resource is a "toolkit" that assists stakeholders with estimating the economic impacts of local food systems. The toolkit was developed by a group of regional economists with financial support from USDA AMS. The modules of the toolkit provide guidance on resources that can be used to frame the assessment process, such as utilizing secondary data and methods for primary data collection, as well as best practice guidance, such as recommending that stakeholders comprise an advisory team of local representatives when undertaking a study. The toolkit modules also explain how to conduct "input-output analysis" to estimate the impacts of local food investments on jobs, income, and gross regional product in a region. AMS's toolkit has been accompanied by an extensive outreach effort.[3]

Analysis of farmers market manager survey data

USDA AMS periodically administers a national survey to all farmers market managers listed in their Farmers Market Directory to collect information on market size, advertising and promotion activities, funding sources, changes in market performance, administrative structure, and community outreach and engagement activities. In 2015, 1396 farmers market managers responded to the survey. Of these responses, 104 of the markets reported receiving 129 FMPP awards. The time period of when the award was received was unspecified in the survey. We use the survey results to make three types of comparisons among FMPP recipients and non-FMPP recipients: the proportion engaged in community outreach and service, change in market performance between 2013 and 2014, and the administrative capability of markets. We interpret these as correlations (and not causal findings) due to the potential for differences that exist among recipient and non-recipient farmers markets that we discussed previously.

We compare the extent to which FMPP recipients are undertaking activities at the market with community benefits relative to non-FMPP recipients in Figure 2. We excluded survey respondents that did not know if their market had received an FMPP award. Offering special events/programming, encouraging healthy diets, offering services to the community, and/or serving as a business incubator were undertaken by 86, 83, 67, and 34%, respectively, of FMPP recipients. These were greater than the corresponding percentages for non-FMPP recipients, which were 73, 60, 46, and 19%, respectively.

The survey also asks managers if various attributes of the market "decreased," "stayed the same," or "increased" between 2013 and 2014. The responses were recorded on a scale from 1 to 3, respectively. The attributes include the level of sales, number of customers, number of repeat customers, number of producers/vendors, vendor retention, vendor recruitment, diversity of products, community/civic contribution and outreach by the market, community support for the farmers market, and financial health of the market. We see in Table 2 that FMPP recipients generally experienced greater increases in market performance between 2013 and 2014 than non-participants. According to market managers, FMPP recipients were

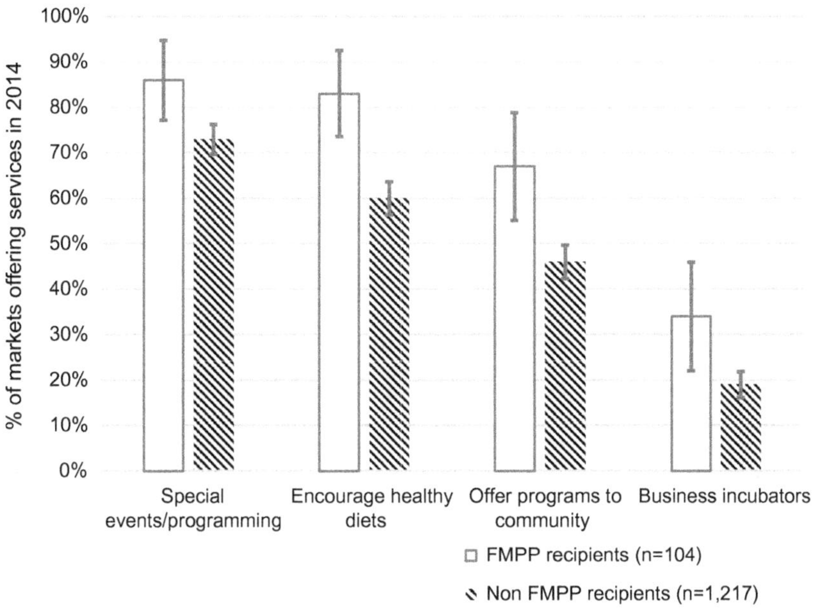

Figure 2. Civic activities offered by farmers markets.
Note: The error bars represent the 99% confidence interval for the respective proportion.

Table 2. Change in market performance between 2013 and 2014.

	Non FMPP recipients	FMPP recipients
Sales**	2.50	2.72
Customers*	2.53	2.66
Repeat customers**	2.58	2.75
Producers/vendors	2.41	2.48
Vendor retention	2.21	2.28
Vendor recruitment**	2.43	2.61
Diversity of products	2.53	2.58
Community/civic contributions/outreach	2.39	2.30
Community support	2.55	2.52
Financial health	2.34	2.42
Responses	*1,217*	*104*

Note: *e* calculated *t* test statistics to compare sample means.
*Statistically significant difference at 5% level; **Statistically significant difference at 1% level.

more likely to experience increases in the level of sales ($p < 0.01$), number of customers ($p < 0.05$), number of repeat customers ($p < 0.01$), and vendor recruitment ($p < 0.01$) than non-recipients. For instance, the average response value of managers that were awarded FMPP awards was 2.72 with regard to whether their farmers market experienced an increase in the level of sales between 2013 and 2014, whereas this value was 2.50 for managers of non-recipient markets.

We further see in Table 3 that FMPP recipients had a greater level of financial resources and viability than non-recipients. In particular, 75% of FMPP recipients had a paid market manager and 39% had paid (non-manager) employees, relative to 51% and 24%, respectively, for non-recipients. Further, the average administrative budget for FMPP recipients was \$59,792, which is greater than the budget of \$21,549 for non-recipient markets ($p < 0.05$).

Table 3. Organizational and demographic characteristics of FMPP recipients vs. non-recipients.

	Non FMPP recipients	FMPP recipients
Paid market manager (%)**	51%	75%
Responses	*1179*	*104*
Paid (non-manager) employees (%)**	24%	39%
Responses	*1168*	*104*
Farmers market budget ($)*	$21,549	$59,792
Responses	*412*	*48*
Metropolitan county*	72%	62%
Responses	*1217*	*104*
2014 county per capita income ($)**	$45,458	$42,042
Responses	*1196*	*102*

Note: We calculated *Z* test statistics in hypothesis tests that compared two sample proportions and t test statistics in hypothesis tests that compared two sample (non-proportion) means.
*Statistically significant difference at 5% level; **Statistically significant difference at 1% level.

These statistics are pertinent since Stephenson et al. (2008) identified low administrative revenue and the dependence on a volunteer manager as two critical factors that are associated with farmers markets that fail.

While farmers markets tend to be located in metropolitan counties since that is where the population is predominately concentrated, we see in Table 3 that farmers markets that were awarded FMPP grants reside in non-metropolitan counties to a relatively greater extent than non-recipient markets. Also, FMPP grants were awarded to markets in counties with a lower per capita income relative to county income levels for markets that did not receive awards (Bureau of Economic Analysis [BEA], 2017). This latter finding could be because FMPP priorities were to support markets in underserved areas and/or because markets in higher income counties may be more self-sustaining with access to other resources for support, which would imply less of a need for FMPP funds.

Case studies of grant recipients

One limitation of the farmers market manager survey is that it is challenging to establish whether an FMPP award caused a change in market performance or whether organizations that were already successful had received awards. Thus, we augment the farmers market manager survey data with brief case studies of a recent FMPP project and LFPP project. Both of the projects featured exceptionally skilled staff and had developed strong community partnerships over the years. While the experience of these projects in implementing a grant would likely vary from organizations with less experience and/or without strong community ties, they are illustrative for several reasons. First, they provide clear examples by which the grant awards made a demonstrable impact and/or catalyzed changes in market activity. Second, they each have well-developed data collection procedures, which allows for the dissemination of a number of metrics pertaining to their projects' impact. Third, they undertook innovative projects, which allows for a more illuminating and nuanced investigation into the challenges that they confronted. Fourth, the projects were geographically proximate to the authors, which allowed for frequent in-person interaction with market organization staff and vendors over a sustained period of time.

Nonetheless, labeling an FMPP and LFPP project unequivocally as a "success" or "failure" can mask the issue that multi-faceted projects can experience both elements, particularly in instances where there are multiple project activities and when there is considerable learning-by-doing associated with implementing innovative projects in low-income communities. The case studies also demonstrate the practical challenges of administering a grant. Impediments to the successful implementation of an FMPP or LFPP award can be attributable to internal organizational factors of award recipients, such as personnel changes or underestimating the challenges associated with implementing a grant, as well as external factors such as inclement weather, adverse changes to economic conditions in the local community, or failure to obtain community support for the project. Identifying and disseminating the lessons learned from FMPP and LFPP projects can be useful to practitioners in implementing local food awards elsewhere.

Thus, we first provide separate overviews of the objectives and outcomes of each project. This is followed by a subsection that contains a more contextual discussion about the unanticipated challenges confronted by the projects, solutions that they developed to overcome them, and lessons learned. We also discuss the extent to which these two case studies are representative of other FMPP and LFPP grant recipients.

FMPP case study: Columbia Heights Community Marketplace (2014)

The Columbia Heights Farmers Market was established in 2010 in a Washington, DC neighborhood with a high level of population density and socioeconomic diversity. The opening of the Saturday morning market was assisted by a 2011 FMPP award that enabled the market's managing organization, the Columbia Heights Community Marketplace (CHCM), to conduct outreach and improve market access to low-income consumers.

CHCM received a second FMPP, at a value of $75,970, in 2014 in order to accomplish several objectives (USDA AMS, 2016c). Many of the project activities that the FMPP supported were implemented in collaboration with local partners and stakeholders. First, CHCM opened a Wednesday evening market from June through September to complement the existing Saturday morning market. At the Wednesday market, CHCM partnered with a healthy school-meal producer to become a DC Free Summer Meals site by providing a snack to children when school is not in session,[4] partnered with a health clinic to assist diabetic patients with attending the market, and offered cooking classes to low-income parents.

CHCM also used the FMPP award to cover the administration costs of a CSA program for SNAP recipients that CHCM would operate on behalf of three of their vendors. CHCM's program allowed SNAP participants to purchase a fruit and vegetable CSA share at a 75% discount. SNAP participants could purchase the shares on a week-to-week basis, which is essential as SNAP benefits cannot be redeemed in traditional CSA programs where the payment occurs up-front in full at the beginning of growing season. Regularly priced CSA shares were also available for purchase by non-SNAP recipients. Further, CHCM partnered with an independent senior care network to deliver CSA shares directly to the residences of low-income senior citizens with limited mobility.

The effect of these initiatives resulted in sales at the Columbia Heights Farmers Market increasing by 12% between 2014 and 2015 to $616,000. Cash expenditures accounted for 77% of the market's 2015 sales, with the remainder occurring from the numerous types of benefits and vouchers that the market accepts. These include SNAP, Women, Infant, and Children (WIC), and Senior Farmers Market Nutrition Program federal food assistance benefits; a $10 "Bonus Bucks" incentive that CHCM provides to recipients to match the redemption of federal benefits; vouchers from the Washington, DC Department of Health's "Produce

Plus Program" that are available to low-income residents; and a "FVRx" program, in which doctors provide weekly vouchers ("prescriptions") to participants to redeem for fruits and vegetables at the market. Columbia Heights Farmers Market has become a prominent farmers market for the redemption of WIC and Produce Plus Program benefits within Washington, DC. Further evidence of the demand for the market's services is that beneficiaries typically line up between 60 and 90 min prior to the market opening to wait for the incentives.

Further, the grant award enabled CHCM to launch a new umbrella organization, Community Foodworks, which assumed management of the Columbia Heights Farmers Market along with other pre-existing farmers markets in the Washington, DC metropolitan area. Community Foodworks was able to use the experience it acquired from administering the Columbia Heights Farmers Market to implement similar initiatives, such as Bonus Bucks, at these other markets.

LFPP case study: Crossroads Community Food Network (2014)

The Crossroads Community Food Network (CCFN) manages the Crossroads Farmers Market in Takoma Park, MD, which is near the "crossroads" of the Montgomery County (MD), Prince George's County (MD), and Washington, DC borders. The market is situated next to a WIC clinic in an ethnically diverse community with high levels of unemployment and food insecurity. In 2007, the Crossroads Farmers Market became one of the first farmers markets in the United States to offer subsidies for the redemption of food assistance benefits.[5] Over time, CCFN perceived that a critical economic development strategy in their community could be the creation of businesses that produce value-added food products.

In 2014, CCFN was awarded a $99,430 LFPP grant to launch a "Microenterprise Training Program" (USDA AMS, 2016d). The first aspect of the program is to assist aspiring food entrepreneurs with creating a business plan, applying for licenses, obtaining insurance, understanding food safety fundamentals, and developing marketing strategies. To convey this knowledge, CCFN administered a five-month bilingual training program that consisted of 10 workshops, optional enrichment classes in partnership with University of Maryland Extension, and individual mentoring. CCFN also partnered with local county health departments to clarify the procedures for obtaining the appropriate licenses and to ensure that forms are translated into accessible languages for community members. Multiple classes were developed and offered during 2015 and 2016. In addition, 254 h of one-on-one technical assistance was provided to trainees outside of the classroom to assist them with completing permit applications, food safety plans, and marketing techniques.

At the completion of the project, 72 potential entrepreneurs graduated from the training program, and 10 had completed business plans. Forty-four trainees had become a certified food manager in their respective county. Upon completion of the training program, a number of graduates have launched their own businesses and begun marketing their products in the region at farmers markets and festivals and through catering. Further, one of the participants began selling a product at a nearby Whole Foods.

A second major component of the LFPP was to open a shared-use commercial kitchen (the Takoma Park Silver Spring Community Kitchen) that will be accessible to training participants by renovating a kitchen at a nearby church. Access to a commercial kitchen is essential for value-added food businesses as only certain food products, such as jellies, hard candy, and baked goods, can be prepared in a home kitchen under Maryland's cottage food law.[6] When the construction of the kitchen experienced delays, CCFN identified other licensed kitchens, including a local Meals on Wheels kitchen, that program graduates could use to avoid delaying the launch of their business. The commercial kitchen did not open during the two-year LFPP grant period, although it is scheduled to open in 2017.

Discussion

One unanticipated challenge that CHCM confronted was in educating SNAP participants about the structure of a subscription-style CSA program. This challenge existed because (a) SNAP recipients did not have previous experience in participating in CSA programs and (b) there was no disincentive for SNAP recipients to not pick up a share since, unlike a traditional CSA, they did not have to pay for their subscription up-front in full. Participation challenges existed, particularly among adults without children and those with high barriers to attending the market every week. While 30 participants in the CSA program fulfilled at least 85% of their pickups, 29 did not, with 11 participants only participating in 15% or fewer of the pickup weeks. CHCM implemented a number of initiatives to maintain participation in the CSA throughout the year. They found that text message reminders were less effective, but that allowing participants to pay for their shares in advance or providing an additional financial pick-up incentive of $8 per bag boosted participation rates.

CCFN also experienced some unanticipated challenges. The large demand for the training program among Hispanic speakers resulted in CCFN developing segregated classes in English and Spanish. While this resulted in less diverse classes than CCFN initially envisioned, it was needed because the substantial time involved with translating in a bilingual classroom was inefficient. Another unanticipated challenge that CCFN confronted was that the immigrant entrepreneurs in their training programs had high level of levels of mistrust and misunderstanding of American financial institutions. Further, developing a business plan was viewed of secondary importance to some of the lower income participants relative to devoting time to generating revenue from selling food products, and some did not have literacy levels needed to write a business plan. As a consequence, 10 business plans were developed instead of 30 that were initially envisioned, and none of the training program participants received financial credit to start a business at the completion of the program. CCFN subsequently partnered with local financial nonprofits to assist training participants on a going-forward basis with the process of establishing a credit history and understanding basic banking services.

To summarize, the case studies demonstrated the importance of two recurring themes that are prominent in the successful implementation of grants: having strong stakeholder support for the project, and partnerships with other organizations in the community. Since both of the case studies are urban-based, some of the project activities (such as delivering CSA shares from a farmers market to a nearby senior center) and implementation challenges (such as language barriers) may not be as relevant for projects in rural-based settings. Nonetheless, the high opportunity costs confronting some of the intended grant beneficiaries impeded their ability to participate fully in the project activities, such as picking up a CSA share every week or writing a business plan. This issue is likely to confront grant recipients irrespective of geographic location, and anticipating and rectifying such challenges can be critical for a project's success.

Conclusion

Farmers markets and local food system venues have been the centers of innovative activities across the United States in the past decade by fostering new supply chains, influencing the shopping habits of consumers, and addressing food system inequities. In this article, we review the history and impacts of two key local food grant programs – FMPP and LFPP. FMPP was one of the first federal programs specifically targeted to local food markets when it was launched in 2006. It has incentivized experimental projects with the aim of providing replicable innovations to be leveraged in other USDA programs and other related initiatives. Since USDA AMS can only fund some of the applications that it receives, it disseminates lessons learned from these projects via reports, blogs, conferences, and other outreach to stakeholders.

Due to the non-standard and non-randomized structure of the programs, we rely upon a farmers market manager's survey, previous literature, and case studies to draw conclusions about their impacts. The farmers market survey data reveals that there is a high association between FMPP recipients with both organizational administrative capacity and improved market performance between 2013 and 2014. Our case studies demonstrate that while many factors are necessary for administering a successful grant, having stakeholder support and partnerships with other community organizations is critical in ensuring that these grants are implemented successfully, with the ultimate goal of being self-sustaining. The case studies also highlight the challenges in operating among low-income communities, where language barriers, cultural differences, lower educational levels, and/or high opportunity costs confronting the project's intended beneficiaries can impede the effectiveness of implementing a grant.

Resources available to guide economic development professionals include formal processes like CEDS and SET, the USDA AMS economic impact toolkit, and resources developed by local food organizations. An example of the latter instance is resources available from the Farmers Market Coalition, a national organization that represents farmers markets, that provide technical assistance and guidance on managing farmers markets.[7] Increases in the size of both FMPP and LFPP grants occurred in 2016. Although there was a profound increase in the capacity for managing local food markets in some parts of the country in the past decade, which is partly attributable to FMPP awards, the future effectiveness of these programs will be influenced by whether there is the organizational capacity to manage larger grants by local food organizations and if these larger grants can support the development of such organizations in regions where they do not currently exist. This latter issue is pertinent since FMPP and LFPP are stipulated to prioritize applicants in areas of concentrated poverty in which organizational capacity may be diminished.

Disclaimer

The views expressed are those of the authors and should not be attributed to the US Department of Agriculture.

Notes

1. The use of FMPP and LFPP award funds is restricted to certain types of costs, which are specified on the AMS website (US Department of Agriculture Agricultural Marketing Service [USDA AMS], 2016a).

2. This assistance included sponsoring 137 in-person grant-writing workshops across the entire United States. In particular, in-person workshops were held in Puerto Rico and the Virgin Islands, as applicants from US territories are eligible for FMPP and LFPP awards. The resources from these grant-writing workshops are available at https://www.amsta.net/.
3. Further details about the toolkit can be accessed at www.localfoodeconomics.com.
4. The DC Free Summer Meals Program reimburses free nutritious meals provided to children 18 and younger in low-income areas when school is out of session in the summer.
5. In 2010, CCFN received an FMPP to launch "Eat Fresh Maryland," which provided support and technical assistance to farmers markets across Maryland in accepting food assistance benefits. The administration of this program was subsequently transferred from CCFN to the Maryland Farmers Market Association.
6. A summary of Maryland's cottage food law is located at https://extension.umd.edu/mredc/specialty-modules/cottage-food-business-law-md.
7. These resources can be accessed at https://farmersmarketcoalition.org/.

Disclosure statement

No potential conflict of interest was reported by the authors.

References

Ahearn, M., & Sterns, J. (2013). Direct-to-consumer sales of farm products: Producers and supply chains in the southeast. *Journal of Agricultural and Applied Economics, 45*, 497–508. doi:10.1017/S1074070800005010

Benson, M., Russell, M., & Kane, D. (2015). *Farm to school grant program fy 2013- fy 2015 summary of awards*. Alexandria, VA: US Department of Agriculture, Food and Nutrition Service.

Bureau of Economic Analysis. (2017). *Regional economic accounts*. Retrieved from https://bea.gov/regional/index.htm

Feenstra, G.W., Lewis, C.C., Hinrichs, C.C., Gillespie, G.W., Jr., & Hilchey, D. (2003). Entrepreneurial outcomes and enterprise size in US retail farmers markets. *American Journal of Alternative Agriculture, 18*, 46–55. doi:10.1079/AJAA200233

Hughes, D.W., & Boys, K.A. (2015). What we know and don't know about the economic development benefits of local food systems. *Choices, 30*, 1–6. Quarter 1.

Hughes, D.W., Brown, C., Miller, S., & McConnell, T. (2008). Evaluating the economic impact of farmers markets using an opportunity cost framework. *Journal of Agricultural and Applied Economics, 40*, 253–265. doi:10.1017/S1074070800028091

Hughes, D.W., & Isengildina-Massa, O. (2015). The economic impact of farmers markets and a state level locally grown campaign. *Food Policy, 54*, 78–84. doi:10.1016/j.foodpol.2015.05.001

Hustedde, R.J., Shaffer, R., & Pulver, G. (2005). *Community economic analysis: A how to manual*. Ames, IA: Iowa State University, North Central Regional Center for Rural Development.

Jablonski, B.B.R., & Schmit, T.M. (2014). *'Local' producers' production functions and their importance in estimating economic impacts*. Ithaca, NY: Cornell University Charles H. Dyson School of Applied Economics and Management. WP 2014–15.

Low, S.A., & Vogel, S. (2011). *Direct and intermediated marketing of local foods in the United States*. Washington, DC: US Department of Agriculture, Economic Research Report number 128.

Miller, S., & Roper, N. (2013). *Farmers Market Promotion Program: Grant activities & impacts 2006–2011*. Miami, FL, New Orleans, LA: Farmers Market Coalition, Market Umbrella.

Onken, K.A., & Bernard, J.C. (2010). Catching the "local" bug: A look at state agricultural marketing programs. *Choices. 25*, 1–7. Quarter 1.

Ragland, E., & Tropp, D. (2009). *USDA national farmers market manager survey 2006*. Washington, DC: US Department of Agriculture, Agricultural Marketing Service.

Schmit, T.M., & Gomez, M.I. (2011). Developing viable farmers markets in rural communities: An investigation of vendor performance using objective and subjective valuations. *Food Policy, 36*, 119–127. doi:10.1016/j.foodpol.2010.10.001

Stephenson, G., Lev, L., & Brewer, L. (2008). 'I'm getting desperate': What we know about farmers markets that fail. *Renewable Agriculture and Food Systems, 23*, 188–199. doi:10.1017/S1742170507002153

Stronger Economies Together. (2016). *Success stories*. Retrieved from https://srdc.msstate.edu/set/current-regions/success-stories

US Department of Agriculture Agricultural Marketing Service. (2016a). *Grants & opportunities*. Retrieved from https://www.ams.usda.gov/services/grants

US Department of Agriculture Agricultural Marketing Service. (2016b). *FMPP final performance reports*. Retrieved from https://www.ams.usda.gov/services/grants/fmpp/reports

US Department of Agriculture Agricultural Marketing Service. (2016c). *FMPP 2014 performance reports:DC-0036 Columbia Heights Community Marketplace $75,970*. Retrieved from https://www.ams.usda.gov/sites/default/files/media/14-FMPPX-DC-0036_Final%20Report.pdf

US Department of Agriculture Agricultural Marketing Service. (2016d). *LFPP 2014 performance reports: MD-0082 Crossroads Community Food Network, Inc. $95,430*. Retrieved from https://www.ams.usda.gov/sites/default/files/media/MD-0082%20Crossroads%20community%20Food%20Network%2C%20Inc.%20%2499%2C430.pdf

US Department of Agriculture Food and Nutrition Service. (2017). *SNAP retailer data*. Retrieved from https://www.fns.usda.gov/snap-retailer-data

US Department of Agriculture National Agricultural Statistics Service. (2016). *2015 local food marketing practices survey*. Retrieved from https://agcensus.usda.gov/Publications/Local_Food/

US Senate Committee on Agriculture, Nutrition, & Forestry (US Senate). (2014). *2014 farm bill*. Retrieved from https://www.ag.senate.gov/issues/farm-bill

The Local Agrifood System Sustainability/Resilience Index (SRI): Constructing a data tool applied to counties in the southern United States

John J. Green, Jim Worstell, and Caroline Canarios

ABSTRACT
Discussions of sustainability, including local agrifood systems research, policy, and practice, increasingly address the concept of resilience. Indicators of resilience need further attention if they are to be useful for community development. This project involved conceptualization, operationalization, and measurement of indicators focused on the resilience of local agrifood systems to construct data-driven tools for theory building and informing development programs and policy, including at the local level. The Local Agrifood Systems Sustainability/Resilience Index was constructed through insights from literature and prior case studies. Building from that work emphasizing the importance of locally modifiable qualities of resilience, this study analyzes publicly available data sources at the county level. Piloted with data from counties in the southern US, findings suggest state and regional differences that may be informative to guide future explanatory research and for development and tracking change as new waves of data become available.

Introduction

This study focuses on efforts to construct a composite measure of local agrifood system sustainability/resilience in the form of a county-level index. Rooted in a critical review and synthesis of ecological resilience literature and building on a series of nine previously conducted qualitative case studies (Worstell & Green, 2017), the specific emphasis in the present research note is on utilizing publicly available secondary data to derive indicators to inform development planning, monitoring, and evaluation processes. Piloted with data from the 13 state southern region as designated by the US Department of Agriculture (USDA), findings suggest important state and regional differences. The index may be used to benchmark contemporary local agrifood system development and monitor change over time as more data become available. These index data, available at the county level, may complement and supplement other efforts providing data at the state, regional, and national levels such as the USDA's Local Foods Marketing Practices Survey (2015).

Conceptual framework

Defining and pursuing sustainability can be seen as a "wicked problem" (Rittel & Webber, 1973; World Bank, 2014) given the difficulty of conceptualization, measurement, and evaluation. As discussed by Worstell and Green (2017), there are researchers who see resilience as an alternative and innovative perspective on sustainability that stresses an observable biological and physical phenomenon. Scholars and practitioners have constructed multiple theoretical frameworks and models to inform work on ecological resilience (Cabell & Oelofse, 2012; Carpenter et al., 2012; Folke et al., 2010; Frankenberger, Mueller, Spangler, & Alexander, 2013; Rockefeller Foundation, 2014; Walker & Salt, 2006). Resilience has been applied in fields as diverse as ecology (Holling, 1973), disaster mitigation (Adger, Hughes, Folke, Carpenter, & Rockström, 2005; Green, Gill, & Kleiner, 2006), community and economic development (Han & Goetz, 2015; Matarrita-Cascante, Trejos, Qin, Joo, & Debner, 2017; Sherrieb, Norris, & Galea, 2010), and social-psychological wellbeing (Schultze-Lutter, Schimmelmann, & Schmidt, 2016), among others.

The concept of ecological resilience emerged from problems confronted in attempts to obtain sustained yields in managed systems, along with observation of adaptive cycles that maintain natural ecosystems' relationships and functions (Holling, 1973). As articulated by Holling and strongly emphasized by Worstell and Green (2017), ecological resilience has an observable biological and physical reality: how much disturbance a system can withstand. Resilient systems last (albeit through ongoing adaptation), while non-resilient systems do not. This conceptualization is widely used, increasingly among climate change analysts (e.g. US Climate Change Science Program [US CCSP], 2008). Numerous studies focus on developing frameworks and theories of resilience and attempt to measure and analyze resilient agricultural systems (Lengnick, 2015). As previously mentioned, resilience thinking has been applied to a range of substantive topics, and collectively, these insights help to connect concepts of resilience to various spatial units of analysis.

Insights from nine case studies of long lasting local agrifood systems in areas of the southern US with few other regional successes enabled integration and synthesis of the above-cited frameworks describing resilient systems into eight qualities that appear to be necessary for resilient systems (Worstell & Green, 2017). Foremost among these is the quality of self-organization. Although numerous features of local agrifood systems are important, self-organization in agrifood systems is manifested in those dimensions that are locally modifiable through collective action. As noted by others (Matarrita-Cascante et al., 2017; Wilson, 2012), it is important to direct attention to people's willingness to take responsibility and control of their development via responsive strategies. Scholars have pointed out that local agrifood systems represent the convergence of community and economic development initiatives (Christensen & Phillips, 2016; Deller, Lamie, & Stickel, 2017). So-called "alternative" and "local" agrifood systems can build bridges between traditional agriculture and more effective natural resource management (Milestad, Westberg, Geber, & Björklund, 2010). Of particular interest is the proposed ability of localized systems to adapt to sudden and external forces. Such forces might result from short-term shocks or long-term stressors ranging from droughts, floods, and climate change to natural disasters and economic downturns. Locally self-organized responses to many such disasters have been shown to be characteristic of resilient communities (Adger et al., 2005; Tidball & Krasny, 2013). Through engaging local people, kindling relationships, and enhancing participation, people working to organize agrifood systems in their communities may be able to help promote inclusiveness, resource mobilization, and knowledge sharing, thereby contributing to broader community development processes. Local agrifood system entities, such as farmers markets, can enhance the adaptive capacity of the systems in which they are embedded to deal with shocks and stresses by collaborative learning between producers and consumers. This adaptive potential arises partially due to close links and face-to-face interactions (Hinrichs, Gillespie, & Feenstra, 2004; Milestad et al., 2010) and the civically engaged nature of these enterprises (Lyson & Guptill, 2004).

This conceptualization of local self-organization does not preclude inclusion of inputs provided from outside the local system when economies of scale or lack of local processing/ manufacturing makes such nonlocal inputs more reliably available at a given point in time. However, the literature cited and the researchers' case studies indicate local self-organization and provision of any such inputs is crucial to resilience. The presence of locally developed organizations and institutions demonstrates collective potential to adapt to change.

Building on previously published materials and their own case studies in Arkansas, Tennessee, and Mississippi, the research team for this article sought to elaborate and augment existing theory to conceptualize, operationalize, and measure local agrifood system sustainability/resilience.[1] Eight qualities (i.e. dimensions) of sustainable and resilient local agrifood systems were identified as common to all nine case studies and consistent with the themes identified by some of the most prominent resilience frameworks (e.g. Cabell & Oelofse, 2012; Carpenter et al., 2012; Frankenberger et al., 2013). From this research, a sustainable and resilient local agrifood system has people who are connected to share information and resources. They are independent yet tightly connected to other communities, markets, and government policies and programs. It also entails self-organization with some of its own initiatives for production, processing, and marketing enterprises, but it embraces a conservative approach to innovation. Such a system maintains some back-ups, redundant units, and pathways, and is able to accumulate resource reserves and physical infrastructure to withstand and/or adapt to disturbance, sometimes going through transformations. Additionally, such agrifood systems entail commitment to ecological integrity and the presence of diverse and complementary enterprises. The labels associated with these dimensions of sustainable and resilient local food systems are listed below, spelling the acronym CLIMATED (Worstell, 2017). The dimensions marked with asterisks are the focus of the remainder of this article.

- Connectivity*
- Local self-organization*
- Innovation
- Maintenance*
- Accumulation
- Transformation
- Ecological integrity*
- Diversity*

Quantitative measurement of these qualities will enable future hypothesis testing about whether these qualities are necessary and sufficient predictors of resilient systems and, if so, providing a tool to help community development practitioners take an informed approach to capacity building and increasing resilience.

Operationalizing the concepts

Beyond representing qualities of the system that local people can influence through their individual and collective actions, in choosing local sustainability/resilience indicators, the research team also directed attention toward those dimensions that could be measured at the county level. This is important for conceptual development and clarity, because any effort to operationalize resilience including broader/higher scale resources, such as programs operated through government agencies and non-governmental organizations, runs the risk of measuring resilience at that larger scale. Although important for broader regional development, these approaches result in measures that can only address the resilience at the higher unit of analysis. Instead, this project sought to identify and measure those qualities that could be self-organized through local initiative in local spaces where people can engage in activities to change their agrifood system through everyday activities and interactions.

Though bonding, bridging, and linking social capital are subsumed as part of the connectivity quality of resilient systems (Worstell & Green, 2017), intentional effort was made not to include more general measures of social capital (Rupasingha, Goetz, & Freshwater, 2006), social resilience (Maclean, Cuthill, & Ross, 2014), socioeconomic position, or health and wellbeing, such as the Social Science Research Council's (n.d.) Measure of America operationalization of the Human Development Index for application to the US and the University of Wisconsin/Robert Wood Johnsons Foundation's County Health Rankings (Remington, Catlin, & Gennuso, 2015). This was not because they are unimportant for understanding resilience, but rather because of the need for conceptual distinction. Scholars, practitioners, and policy-makers should be interested in whether there are relations between local agrifood system sustainability/resilience and other socioeconomic, health, and wellbeing indicators. For instance, Deller, Canto, and Brown (2017) investigated the relationship between access to local foods and measures of community health. Mixing social, health, and wellbeing measures within the index presented in this article would prevent research similar to theirs, thereby reducing analytic options for testing important associations.

The research team chose indicators from publicly accessible databases to reflect the qualities of sustainability and resilience identified through the previously mentioned case studies of local agrifood systems in the southern US. Although researchers were not able to identify county-level comparable data for all eight of the dimensions of sustainable/resilient local agrifood systems as presented in the CLIMATED framework, they constructed indicators for five of the dimensions that could be combined to form a composite Local Agrifood System Sustainability/Resilience Index (SRI). They did this drawing primarily from the 2012 and 2007 National Census of Agriculture, with additional data from the USDA's 2013 Food Atlas, 2013 Farm to School Census, and state-level regulatory records assembled by the research team during 2014 (Table 1). (Future iterations of the SRI may include additional dimensions of the CLIMATED framework and will make use of updated data as appropriate.) In choosing variables, they privileged data available for most counties in each of the states. The dimensions/qualities and associated indicators are discussed below.

Connectivity

Only one indicator was available in county-level databases that addressed connectivity. This was the percent of farm operations with internet access.

Table 1. Local Agrifood System Sustainability/Resilience Index (SRI) indicators and data sources.

Dimension	Indicators
Connectivity	Farms with internet % farm operations with Internet access (Source: 2012 Census of Agriculture)
Local self-organization	Farmer lives on farm % principal operators living on farm Farms with local food alternatives scale (α = .739) % operations with on-farm packing % operations with direct marketing to retail % operations with community supported agriculture Community local food alternatives index Cumulative score on having meat processing facilities, farmers markets, and farm to school programs (Sources: 2012 Census of Agriculture; 2013 Food Environment Atlas; 2013 Farm to School Census; 2014 online research of state regulations and listings)
Maintenance	Average age of farm operator Each county as a % of the highest average age in the region (68.7) Reverse coded for lower average ages to have higher scores on final indicator Stability-change % change in number of farm operations between 2007 and 2012 (Sources: 2007 and 2012 Census of Agriculture)
Ecological integrity	Low chemical input scale (α = .759) % agricultural land not treated with herbicides % agricultural land not treated with insecticides *(Note: Acres of "crop land" in the denominator. Numerator for insecticides excludes treatment for nematodes.)* Organic farms % operations certified organic Management intensive livestock % operations practicing management intensive/rotational grazing (Source: 2012 Census of Agriculture)
Diversity	Production diversity index Row crop diversity Average % of operations producing across seven different row crop options *(Note: Does not indicate these crops were grown on the same farms.)* Vegetable production % operations with vegetables harvested Livestock production % operations with livestock sales (Source: 2012 Census of Agriculture)
Local Agrifood System Sustainability/Resilience Index	All indicators listed above were standardized and then coded into a summative index.

Locally self-organized

Data on three variables were used to calculate scores for the concept of a system's level of being locally self-organized. Local farm management was measured by one variable: percent of principal operators living on their farm. This is assumed to reflect degree of locally self-organized management in contrast to absentee ownership. The indicator to measure prevalence of farm operations with local food alternatives was composed of three variables (α = .739): percent of operations with on-farm packing, percent of operations with direct marketing to retail, and percent of operations with community supported agriculture (CSA).[2] Locally organized community food alternatives consisted of a cumulative score of individual binary variables representing whether farmers markets and farm to school programs were present in the county and whether meat-processing facilities were present within the county or in adjoining counties.

Maintenance

This quality was measured with two indicators. The first variable was average age of farm operator. Each county was given a score of the percent of the highest average farm operator age in the region (68.7). These scores were then reverse coded for lower average ages to result in higher scores on the final indicator based on the assumption that younger farmers provide the potential for greater redundancy of farm operators over time. The second variable was the percent decrease or increase in the number of farm operations between 2007 and 2012. Scores were reverse coded such that counties with lower percent decreases (and increases) had higher scores.

Diversity

Scores for this quality reflect production diversity. Diversity at the county level was derived from three indicators: row crop diversity (average percent of operations producing across seven different row crop options), vegetable production (percent of operations with vegetables harvested), and livestock production (percent of operations with livestock sales).

Ecological integrity

Researchers used data from four variables to calculate ecological integrity scores. The low chemical input index ($\alpha = .759$) was created from two variables: percent of agricultural land not treated with herbicides, and percent of agricultural land not treated with insecticides. Acres of crop land were used as the denominator in these calculations. The numerator for insecticides excludes treatment for nematodes. The two other variables included for the ecological integration quality were organic practices (percent of operations certified organic) and the ecologically integrated practice of management intensive/rotational grazing (percent of operations practicing management intensive/rotational grazing).

To reiterate, during the conduct of this study, the research team did not identify useful variables at the county level for the other three of the eight qualities of resilience: innovation, maintenance, and transformation. Effort was made to do so using available data, but when presented across many of the 15 meetings (described below), the indicators were not seen by local food system development practitioners as being useful or informative. This is an area requiring more work.

Analysis and computation of the Local Agrifood System SRI

Counties/county equivalents were the units of analysis, focusing in this project on the 13 state southern region as designated by the USDA: Alabama, Arkansas, Florida, Georgia, Kentucky, Louisiana, Mississippi, North Carolina, Oklahoma, South Carolina, Tennessee, Texas, and Virginia. Starting with a total of 1344 counties, 43 were treated as missing due to insufficient data on some key variables, especially number of farm operations in either 2007 or 2012. All other missing values in the index were counted as 0 on the basis that the practices under consideration were not prevalent enough to avoid data suppression. Final analysis was based on 1301 cases (counties).

All percentages were transformed to *z*-scores for the purpose of normalization. Summated multi-variable scales (farms with local food alternatives and low chemical input) and indices (community food alternatives and production diversity) were calculated by summing *z*-scores across component variables. *Z*-scores were then ranked (higher scores have higher ranks) by county across the entire region and categorized by quartile groupings. Because of low organic certification rates, this variable was only ranked in two categories.

The research team also created a composite "Local Agrifood System Sustainability/ Resilience Index" (SRI), summing and standardizing the standardized scores across all of the separate indicators. Both unweighted and weighted versions of the composite index were calculated. The latter was done using principal components analysis (Organization for Economic Co-Operation and Development [OECD], 2008), applying the loadings on factor one as the weights for each variable. This is discussed in further detail in the following section. Since the indicators were standardized and ranks were calculated across the southern region (rather than within each state separately), findings show the position of each county within the entire region. (Future analysis will feature both state and regional rankings for each county.)

Following initial construction of the SRI, tables and maps were prepared and shared at 15 meetings across local, regional, national, and international settings with participants who included farmers, extension professionals, policy advocates, and other researchers.[3] Besides outreach and dissemination of preliminary findings, a major purpose of those presentations and discussions was to address the validity of the index and interpretation of meaning. In three meetings held in the southern region, attendees helped to "workshop" the SRI and provided feedback on the relevance of the measures for informing local initiatives. After each presentation, the SRI was re-evaluated and sometimes modified to arrive at the version presented in this article.

Findings

It is critically important for readers to note that the SRI was not constructed as a "scale" where multiple indicators would be expected to be highly inter-correlated, each one measuring relatively the same concept and thereby serving as indicators of some broader latent construct with internal reliability. Instead, this "composite index" combined a range of indicators that were: (a) not expected to be highly inter-correlated, and (b) better thought of as causes of or contributors to our latent construct, that is "Local Agrifood System Sustainability/ Resilience." The methodological approach used to construct this composite index was somewhat similar to the approach used for the County Health Rankings System (Remington et al., 2015), with the exception of the SRI being based on regional versus statewide rankings.

There is considerable debate concerning whether individual items included in composite indices should be weighted equally or assigned weights based on theoretical and/or statistical analysis (Organization for Economic Co-Operation and Development [OECD], 2008). Equal weighting was originally used for the analysis, because it is likely more intuitive for general users and is theoretically parsimonious (Stapleton & Garrod, 2007). However, it is critically important to note that the number of indicators from each dimension or quality included in computation of the composite index essentially provide de facto weights. For instance, the resilience quality of a system having locally self-organized features includes three indicators, as does the quality ecological integration. However, the quality connectivity only has one indicator. Since the overall composite SRI was computed through summation of all indicators across the qualities, local self-organization and ecological integration are more heavily represented than connectivity.

Recognizing that some analysts criticize equal weights in composite indices, including peer-reviewers of a previous version of this article, and following the recommendation of the OECD (2008) to assess the impact of different weighting schemes, the research team also used principal component factor analysis (PCA) to determine if alternative weighting procedures would have led to notably different results. Four factors were extracted explaining 60% of variation. The loading scores on the first principle component were used as weights in calculating an alternative form of the SRI. To test the potential difference between the use of equal weights and the PCA method, the correlation between the two versions of SRI was calculated (Pearson's $r = .907$ and Spearman's $\rho = .898$, $n = 1301$ counties).

Both forms of the SRI were used to test for state level patterns. The descriptive findings presented here identify patterns within and between states on the individual indicators and the overall unweighted and weighted SRI. State-level patterns were identified for these indicators (Table 2), and regional concentrations (within and across state lines) were observed (Maps 1(a) and 1(b)). There were some slight shifts in which states had the highest percentage of counties in the highest quartiles, but the overall rank-order correlation was strong (Spearman's $\rho = .841$, $n = 13$ states). Furthermore, the same states were in the bottom three for lowest percent of counties in the highest quartile (TX, MS, and AL) and in the top three for highest percent of counties in the highest quartile (VA, NC, and KY).

Analyzing the percent of counties within each state scoring within the top quartile when ranked to the region, interesting patterns emerged. While some variation was identified between states concerning whether their counties ranked high on the various qualities, there were major distinguishing patterns between states concerning total composite SRI. Virginia, North Carolina, and Kentucky each had more than 50% of their counties (excluding missing cases) in the highest quartile for the region, thereby representing the greatest levels of local agrifood system sustainability/resilience in the southern region. The maps help to visualize these state-level differences, but they also demonstrate that there are regional clusters that cross state lines. More densely populated and urbanized areas have higher SRI scores, and both west Texas and the multi-state Mississippi Delta region have some of the lowest. State comparisons indicate that the Appalachian region has higher scores than might be expected given previously documented socioeconomic conditions. Although not explored here, these patterns might result from a combination of different agronomic conditions, development legacies in agricultural production, and local political–economic systems, all warranting significantly more research. Determining the reasons for these patterns may prove particularly useful for development practitioners and policy-makers.

Explaining the statistical associations between the SRI and other socioeconomic conditions goes beyond the purpose and scope of this article. However, as a starting point, the research team explored the rank correlations with other important measures of development for future analysis (Table 3). In general, the patterns meet what we might expect. Counties with a higher percent of people living in urban areas in 2010 (weak correlation) and a lower percent of families living below poverty according to 2010 and 2015 US Census Bureau American Community Survey five-year estimates (moderate correlation) also have higher SRI scores. The exception to what might be expected is the weak yet negative correlation between the 2014 version of the Rupasingha et al. (2006) social capital index and SRI.

Table 2. Southern US state comparisons on the Local Agrifood System Sustainability/Resilience Index (SRI).

	Connec-tivity	Local self-organization			Maintenance		Ecological integrity			Diversity		
	Farms w/ internet access	Farmer lives on farm	Farms w/ local food alternatives	Community local food alternatives	Age redun-dancy	Stability/ positive change in # of farms	Low chemi-cal input	Organic farms	Management intensive livestock	Production diversity	Full Index	PCA Weighted-Index
State	Percent of counties in state within the highest quartile for the southern region											
Alabama (67)	9.0	16.4	10.4	29.9	13.4	9.0	13.4	11.9	4.5	19.4	3.0	9.0
Arkansas (75)	16.0	49.3	2.7	13.3	48.0	17.3	13.3	16.0	29.3	22.7	17.3	30.7
Florida (67)	46.3	23.9	47.8	31.3	14.9	46.3	11.9	55.2	11.9	6.0	34.3	13.4
Georgia (48)	40.3	30.2	25.8	17.6	17.6	16.4	14.5	21.4	14.5	25.8	16.4	17.6
Kentucky (120)	15.8	30.8	30.0	12.5	58.3	9.2	60.0	25.0	41.7	44.2	50.0	49.2
Louisiana (62)	29.0	19.4	29.0	8.1	37.1	33.9	16.1	9.7	19.4	12.9	16.1	14.5
Mississippi (82)	7.3	17.1	8.5	9.8	9.8	14.6	14.6	6.1	6.1	13.4	3.7	7.3
North Carolina (100)	40.0	24.0	75.0	46.0	26.0	31.0	11.0	53.0	24.0	83.0	56.0	47.0
Oklahoma (77)	18.2	39.0	2.6	23.4	37.7	7.8	6.5	20.8	7.8	0.0	13.0	22.1
South Carolina (46)	15.2	21.7	19.6	47.8	10.9	52.2	15.2	39.1	10.9	26.1	32.6	32.6
Tennessee (95)	7.4	37.9	17.9	14.7	23.2	8.4	55.8	20.0	37.9	21.1	21.1	31.6
Texas (254)	23.2	5.5	5.5	13.8	12.2	40.2	22.8	20.5	30.3	5.1	8.3	6.3
Virginia (97)	43.4	37.1	66.3	37.1	22.7	35.1	48.5	36.1	55.7	51.5	68.0	61.9

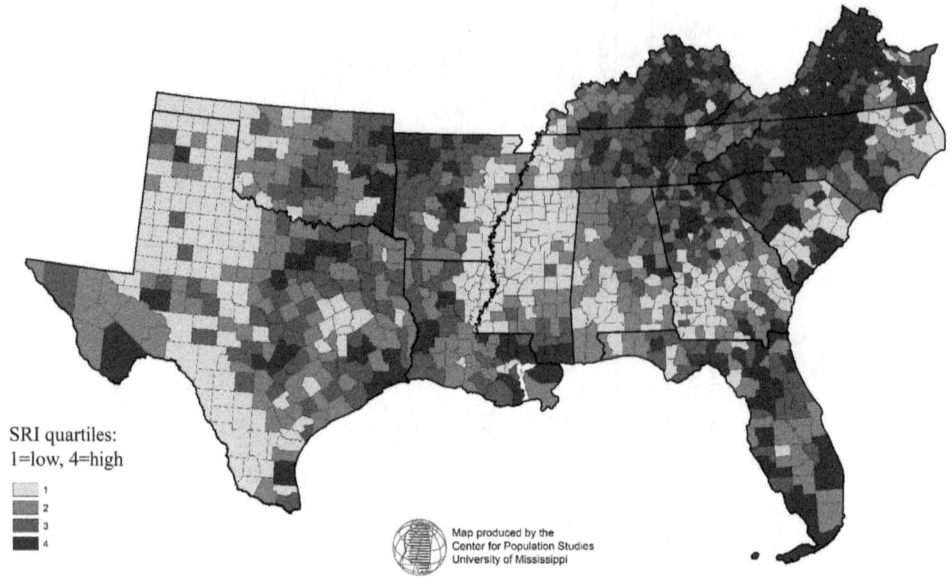

Map 1a. Unweighted composite Local Agrifood System Sustainability/Resilience Index (SRI) in the southern US (quartile ranks).

Notes: Based on summated and standardized scores across 10 indicators, and weighted using principal components analysis. A higher quartile ranking indicates a higher level of sustainability/resilience on this composite index. Data sources: 2012 Census of Agriculture, 2013 Food Environment Atlas, 2013 Farm to School Census, and 2014 review of state policies and regulations; extra calculations by the authors and research team. Analysis based on 1301 counties.

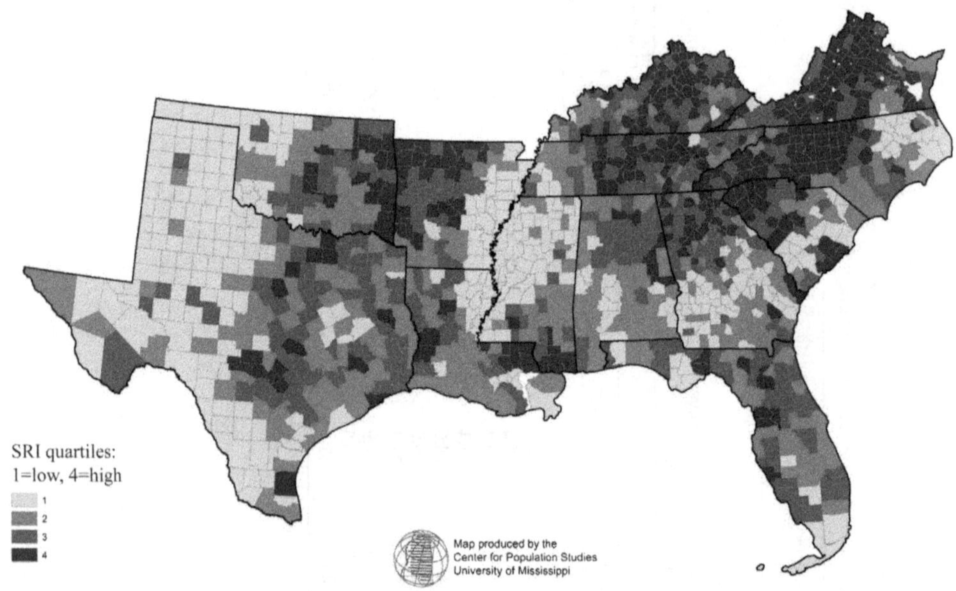

Map 1b. Weighted composite Local Agrifood System Sustainability/Resilience Index (SRI) in the southern US (quartile ranks).

Notes: Based on summated and standardized scores across 10 indicators, and weighted using principal components analysis. A higher quartile ranking indicates a higher level of sustainability/resilience on this composite index. Data sources: 2012 Census of Agriculture, 2013 Food Environment Atlas, 2013 Farm to School Census, and 2014 review of state policies and regulations; extra calculations by the authors and research team. Analysis based on 1301 counties.

Table 3. Spearman's rho (rank correlation) between Local Agrifood System Sustainability/Resilience Index (SRI) and various demographic and development variables (1301 southern US counties).

Indicators	Unweighted Resilience Index	PCA Weighted Index
Percent of families below poverty (2015 5-year American Community Survey Estimates)	−.362	−.359
Percent of families below poverty (2015 5-year American Community Survey Estimates)	−.275	−.263
Social capital (2014, updates of Rupasingha, Goetz, & Freshwater, 2006)	−.116	−.113
Components of social capital		
Associations per 1000 population (2014)	−.119	−.176
Percent voter turnout (2012)	.102	.103
Response rate to 2010 Census (2010)	.441	.467
Nonprofit organizations (2014)	.361	.409

Data sources: SRI (2012 Census of Agriculture, 2013 Food Environment Atlas, 2013 Farm to School Census, and 2014 review of state policies and regulations; extra calculations by the authors and research team); US Census Bureau, American Community Survey 2010 and 2015 5-year estimates (https://factfinder.census.gov/); Rupasingha et al. (2006 with 2014 updates). Analysis based on 1301 counties. For all correlations $p \leq .01$.

However, when investigating the associations between the component scores for measures of social capital and SRI, only the number of associations (e.g. religious, business, sport) per 1000 population had a negative correlation. All of the other components were positively associated with SRI.

Discussion and next steps

Identifying patterns of local agrifood system sustainability and resilience in the southern US may be useful for informing extension, program development, policy innovations, and investment decisions to address sustainability, food security, and health and wellbeing. For instance, the SRI helps to identify specific counties and regions that are doing relatively well in the search for greater resilience, and they can be studied to learn about effective strategies at the local and statewide levels. Those places that are not doing as well might be the focus of additional efforts to spur more local initiatives to create agrifood systems with greater agility and adaptability. This attempt to operationalize resilience at the county level in the US should not be regarded as definitive in regions such as the High Plains, the Delta, and Appalachia, as noted above.[4] More research is needed to ground-truth, authenticate, and refine the index. Particularly useful will be studies relating the index to performance of local agrifood systems – such as claims for crop insurance and survival of farms.

The SRI is based on publicly accessible data sources that are regularly updated. Researchers used 2012 Census of Agriculture data along with other sources (Farm to School Census, Food Environment Atlas) from near that time for this article. At the time of this writing, the 2017 Census of Agriculture was being conducted, so data for comparison will soon be available. At that point, researchers will be able to track temporal change, a key feature of sustainability and resilience missing from the present analysis.

Piloted here for the southern US, the SRI could be measured for counties across the country, and comparable variables could be identified for other nations. As previously noted, the SRI was standardized to the region as a whole for this article. This could be done for the nation as well. In either case, practitioners, development funders, and policy-makers would also benefit from having access to within state rankings (as has been done for the County

Health Rankings program, see: Remington et al., 2015). This is the next step for the research team.

Finally, the SRI may be useful in analyses concerning factors associated with broader socioeconomic wellbeing and health outcomes. For instance, many studies focused on food insecurity use indicators such as the number of food retailers per capita or whether a county is considered a "food desert" because of the lack of nearby grocery stores and supermarkets (Ver Ploeg et al., 2009). Although informative, decisions about the location of these food sources are often far beyond local control (with the exception of tax and regulatory incentives/disincentives). Government programs do exist to spur local food and farmers market activities (O'Hara & Coleman, 2017), and the SRI might be informative for decision-making about what places might be well positioned for leveraging additional resources. The SRI contributes individual measures and a composite index that represents factors that can be mobilized and organized at the local level. These factors may be influenced by broader policies and access to resources such as grants from government agencies and foundations, but they are not fully dependent upon them. They are, however, dependent on some level of local action. Thus, authors of the SRI hypothesize that it will be particularly informative to the field of community development. In this vein, their current work is investigating the inter-relationships between the SRI, measures of human development, and population health.

Notes

1. The terms sustainability and resilience are used together here to connect these two bodies of literature and fields of practice, rather than treating them separately.
2. As noted by one of the peer-reviewers, there are concerns about the accuracy of CSA counts from the Census of Agriculture (for example, see Galt, 2011). The SRI research team acknowledged these challenges, but decided to include the existence of CSAs in a county in the index using Census of Agriculture numbers because this source is still the most comparable across states. It should be noted that the presence of one or more CSA in a county was given a score of one for the community alternatives index.
3. Berry Center Conference, Healthy Foods, Local Farms, Louisville, KY 2014; Food Summit, Mississippi Sustainable Agriculture Network, Jackson, MS 2014 and 2016; National Sustainable Agriculture Coalition Meeting, Big Sky, MT 2014; Small Farm Conference, Kentucky State University, Frankfort, KY 2014; Rural Sociological Society Conference, Madison, WI 2015; Small Farmers Conference, Mississippi Association of Cooperatives and Alcorn State University, Hattiesburg, MS 2015; Southern Sustainable Agriculture Working Group, Mobile, AL 2015; National Sustainable Agriculture Research and Education (SARE) Operating Council, Washington DC 2015; National Sustainable Agriculture Coalition, Jekyll Island, GA 2015; Southern SARE Administrative Council, Atlanta, GA 2015; Western SARE Professional Development Program Conclave, Durango, CO 2015; Association of Natural Resource Extension Professionals and National Association of Community Development Extension Professionals Conference, Burlington, VT 2016; World Congress of Rural Sociology, International Rural Sociology Association, Toronto, Ontario 2016; National Sustainable Agriculture Coalition, Louisville, KY 2017.
4. Regional and state patterns need further attention. For instance, correlation analysis conducted for SRI and percent of families below poverty between counties in Kentucky and those in other states showed different effect sizes. Outside of Kentucky, there was a moderate strength correlation between weighted SRI and percent of families below poverty (2010 poverty $\rho = -.412$, 2015 poverty $\rho = -.299$, $n = 1181$, $p \leq .01$), while for Kentucky the correlation was weaker (2010 poverty $\rho = -.112$, 2015 poverty $\rho = -.202$, $n = 120$, $p \leq .01$).

Acknowledgements

This article is based on work that was supported by the National Institute of Food and Agriculture, US Department of Agriculture (USDA), through the Southern Region Sustainable Agriculture Research and Education (SARE) program. USDA is an equal opportunity employer and service provider. Any opinions, findings, conclusions, or recommendations expressed in this publication are those of the authors and do not necessarily reflect the views of the USDA. The authors would like to thank the numerous staff and students who contributed to the research team for this initiative over the years.

Disclosure statement

No potential conflict of interest was reported by the authors.

Funding

This work was supported by the US Department of Agriculture, Southern Sustainable Agriculture Research and Education Program [sub-award number RD309-134/S001078].

References

Adger, W.N., Hughes, T.P., Folke, C., Carpenter, S.R., & Rockström, J. (2005). Social-ecological resilience to coastal disasters. *Science, 309*, 1036–1039. doi:10.1126/science.1112122

Cabell, J.F., & Oelofse, M. (2012). An indicator framework for assessing agroecosystem resilience. *Ecology and Society, 17*, 18. doi:10.5751/ES-04666-170118

Carpenter, S.R., Arrow, K.J., Barrett, S., Biggs, R., Brock, W.A., Crépin, A.-S., & de Zeeuw, A.D. (2012). General resilience to cope with extreme events. *Sustainability, 4*, 3248–3259. doi:10.3390/su4123248

Christensen, B., & Phillips, R. (2016). Local food systems and community economic development through the lens of theory. *Community Development, 47*, 638–651. doi:10.1080/15575330.2016.1214609

Climate Change Science Program [US CCSP]. (2008). *Our changing planet: The US climate change science program for fiscal year 2009*. Washington, DC: Author. Retrieved from https://www.globalchange.gov/browse/reports/our-changing-planet-us-climate-change-science-program-fiscal-year-2009

Deller, S., Canto, A., & Brown, L. (2017). Food access, local foods, and community health. *Community Development, 48*(5), this issue.

Deller, S.C., Lamie, D., & Stickel, M. (2017). Local food systems and community economic development. *Community Development, 48*(5), this issue.

Folke, C., Carpenter, S.R., Walker, B., Scheffer, M., Chapin, T., & Rockström, J. (2010). Resilience thinking: Integrating resilience, adaptability and transformability. *Ecology and Society, 15*(4), 20. Retrieved from http://www.ecologyandsociety.org/vol15/iss4/art20/

Frankenberger, T., Mueller, M., Spangler, T., & Alexander, S. (2013). *Community resilience: Conceptual framework and measurement Feed the Future learning agenda*. US Agency for International Development. Rockville, Maryland: Westat. Retrieved from https://agrilinks.org/sites/default/files/resource/files/FTF%20Learning_Agenda_Community_Resilience_Oct%202013.pdf

Galt, R.E. (2011). Counting and mapping community supported agriculture (CSA) in the United States and California: Contributions from critical cartography/GIS. *ACME, 10*, 131–162. Retrieved from http://www.caff.org/wp-content/uploads/2012/06/Counting-and-Mapping-CSA-in-the-US-and-CA.pdf

Green, J.J., Gill, D.A., & Kleiner, A.M. (2006). From vulnerability to resiliency: Assessing impacts and responses to disaster. *Southern Rural Sociology, 21*, 89–99.

Han, Y., & Goetz, S.J. (2015). The economic resilience of US counties during the Great Recession. *The Review of Regional Studies, 45*, 131–149. Retrieved from http://journal.srsa.org/ojs/index.php/RRS/article/view/45.2.2/pdf

Hinrichs, C.C., Gillespie, G., & Feenstra, G.W. (2004). Social learning and innovation at retail farmers' markets. *Rural Sociology, 69*, 31–58. doi:10.1526/003601104322919892

Holling, C.S. (1973). Resilience and stability of ecological systems. *Annual Review of Ecology and Systematics, 4*(1), 1–23. doi:10.1146/annurev.es.04.110173.000245

Lengnick, L. (2015). *Resilient agriculture: Cultivating food systems for a changing climate.* Gabriola Island: New Society Publishers.

Lyson, T.A., & Guptill, A. (2004). Commodity agriculture, civic agriculture, and the future of US farming. *Rural Sociology, 69*, 370–385. doi:10.1526/0036011041730464

Maclean, K., Cuthill, M., & Ross, H. (2014). Six attributes of social resilience. *Journal of Environmental Planning and Management, 57*, 144–156. doi:10.1080/09640568.2013.763774

Matarrita-Cascante, D., Trejos, B., Qin, H., Joo, D., & Debner, S. (2017). Conceptualizing community resilience: Revisiting conceptual distinctions. *Community Development, 48*, 105–123. doi:10.1080/15575330.2016.1248458

Milestad, R., Westberg, L., Geber, U., & Björklund, J. (2010). Enhancing adaptive capacity in food systems: Learning at farmers' markets in Sweden. *Ecology and Society, 15*, 29. Retrieved from http://www.ecologyandsociety.org/vol15/iss3/art29/

O'Hara, J.K., & Coleman, C. (2017). The impacts of the farmers market and local food promotion programs. *Community Development, 48*(5), 1–16.

Organization for Economic Co-operation and Development. (2008). *Handbook on constructing composite indicators: Methodology and user guide.* Paris: Author. Retrieved from: www.oecd.org/publishing/corrigenda

Remington, P.L., Catlin, B.B., & Gennuso, K.P. (2015). The county health rankings: Rationale and methods. *Population Health Metrics, 13*(1), 11. doi:10.1186/s12963-015-0044-2

Rittel, H.W., & Webber, M.M. (1973). Dilemmas in a general theory of planning. *Policy Sciences, 4*, 155–169. doi:10.1007/BF01405730

Rockefeller Foundation. (2014). *City resilience framework.* Retrieved from https://www.rockefellerfoundation.org/app/uploads/City-Resilience-Framework1.pdf

Rupasingha, A., Goetz, S.J., & Freshwater, D. (2006, with 2014 updates). The production of social capital in US counties. *The Journal of Socio-Economics, 35*, 83–101. doi:10.1016/j.socec.2005.11.001

Schultze-Lutter, F., Schimmelmann, B.G., & Schmidt, S.J. (2016). Resilience, risk, mental health and well-being: Associations and conceptual differences. *European Child & Adolescent Psychiatry, 25*, 459–466. doi:10.1007/s00787-016-0851-4

Sherrieb, K., Norris, F.H., & Galea, S. (2010). Measuring capacities for community resilience. *Social Indicators Research, 99*, 227–247. doi:10.1007/s11205-010-9576-9

Social Science Research Council. (n.d.). Measure of America. Retrieved July 5, 2017 from https://www.measureofamerica.org

Stapleton, L.M., & Garrod, G.D. (2007). Keeping things simple: Why the human development index should not diverge from its equal weights assumption. *Social Indicators Research, 84*, 179–188. doi:10.1007/s11205-006-9081-3

Tidball, K.G., & Krasny, M.E. (Eds.). (2013). *Greening in the red zone: Disaster, resilience, and community greening* (2014th ed.). Dordrecht: Springer.

Ver Ploeg, M., Breneman, V., Farrigan, T., Hamrick, K., Hopkins, D., Kaufman, P., … Tuckermanty, E. (2009). *Access to affordable and nutritious food-measuring and understanding food deserts and their consequences: Report to Congress.* Washington, DC: US Department of Agriculture Economic Research Service. Retrieved from https://www.ers.usda.gov/publications/pub-details/?pubid=42729

Walker, B., & Salt, D. (2006). *Resilience thinking: Sustaining ecosystems and people in a changing world.* Washington, DC: Island Press.

Wilson, G.A. (2012). Community resilience, globalization, and transitional pathways of decision-making. *Geoforum, 43*, 1218–1231. doi:10.1016/j.geoforum.2012.03.008

World Bank. (2014). *A wicked problem: Controlling global climate change.* Washington, DC: The World Bank. Retrieved from https://www.worldbank.org/en/news/feature/2014/09/30/a-wicked-problem controlling-global-climate-change

Worstell, J. (2017). CLIMATED. *Resilience Blog*, May 18 entry. Retrieved from https://meadowcreekvalley.wordpress.com/category/blog/

Worstell, J., & Green, J. (2017). Eight qualities of resilient food systems: Toward a sustainability/resilience index. *Journal of Agriculture, Food Systems, and Community Development.* Advance online publication. doi:10.5304/jafscd.2017.073.001

Index

Note: **Bold** page numbers indicate tables; *italic* numbers indicate figures.

academic work on local food systems *see* research on local food systems
access to local foods, public health and: amount spent on local foods 65; causation 66; complexity of issues 53–54; context for 49–50; control variables 56–58; data availability 55; defining 'local' foods 55; diet-related disease and health 52–53; direct connection with farmers 52; environment, local food 51, 53; farm to school programs 52; fruit and vegetable consumption 52; incentive programs 65–66; literature review 51–53; local foods, measurement of 55–56; Local Foods Act of 2013 (Ontario, Canada) 50; measurement 54–56; measurement index as research objective 51; modelling approach and methods 53–58; modest positive impact 65, 66; multiple issues addressed by local foods 50; necessary not sufficient, nutritious food as 50; obesity and diabetes, relationship with 53; premature mortality 54; principal components analysis 55, 56; promotion of 49–50; public health, measurement of 54–55; research gap regarding 50–51; results of research 58, **59**, 60, 61, 62, 62–65, *63*; systems thinking 54; WIC Farmers Market Nutrition Program 52
Adalja, A. 13
Adams, D.C. 13
Arku, G. 11–12, 19
authenticity 14
Azuma, A.M. 52

Ballingall, J. 15
Bedimo-Rung, A.L. 58
behavioral attributes, demand for local food and 12
Berning, J.P. 52, 54, 65
Born, B. 21, 50
Boys, K.A. 12, 16
branding schemes, defining 'local' and 8
Brewer, L. 75
Brislen, L. 19
Brown, A. 33
Brown, L. 18, 65, 92
Brunekreef, B. 56–57

Canto, A. 18, 65, 92
Carpio, C.E. 12
Carson, R.A. 18
Casey, M.M. 64
causality, research on local food systems 18 and
Chambers, K.J. 55
characteristics of location, demand for local food and 12
characterization of local foods **9**, 9–11, *10*
Churchill, S.A. 57
civic agriculture 8, 13
Clark, J.K. 16
Cohen, D.A. 58
Colpaart, A. 65
Columbia Heights Community Marketplace 83–84, 85
Community Capitals Framework (CCF) 32
community economic development: consumer demand for local foods 11–14; food security, local foods and 18–19; future growth/potential of local foods 15; local foods and 14–16; opportunity costs in switch to local foods 14–15; producers, impact of local foods on 16–17; public health, local foods and 17–18; quantitative index of local foods **9**, 9–11, *10*; scalability/replicability of local foods models 16; social capital, local foods and 19–21
Community Economic Development Strategy (CEDS) 79
consumer demand for local foods 11–14
consumption, style of, defining 'local' and 7
County Health Rankings and Roadmaps program 54
Crossroads Community Food Network 84–85
cultural traditions, food and 1

Davis, A. 13
Davis, G.A. 16

Deller, S.C. 18, 65, 92
Delphi method, use of for research 33–34
demand for local foods 11–14
Derkatch, C. 50, 65
deserts, food 18–19
diabetes, local food environment and 53
diet-related disease and health 52–53
dot poster surveys 35–36
Dunn, B. 50
Dunne, J.B. 55
Durham, C.A. 55

economic development of communities: Community Economic Development Strategy (CEDS) 79; consumer demand for local foods 11–14; food security, local foods and 18–19; future growth/potential of local foods 15; local foods and 14–16, 74–75; opportunity costs in switch to local foods 14–15; planning resources 79–80; producers, impact of local foods on 16–17; public health, local foods and 17–18; quantitative index of local foods **9**, 9–11, *10*; scalability/replicability of local foods models 16; social capital, local foods and 19–21; Stronger Economies Together (SET) 79–80; toolkit for estimating impact of local foods 80; *see also* rural economic development through local foods
Emery, M. 32–33
environment, local food 51–52, 53
experience, farmers markets as 14
exports from low-income countries 15
extended social accounting matrix (SAM) model 33

failed initiatives as overlooked in research 22
farm to school programs 5, 52
farmers: impact of local foods on 16–17; motivation for 16–17; support programs 17
Farmers Market Coalition 86
Farmers Market Promotion Program (FMPP): activities supported 76–77; applications to 77; awards made 76; background 76–77, **77;** case study of grant recipients 82–84, 86; Columbia Heights Community Marketplace 83–84, 85; dollar value of grants *77*; EBT machines, installation of 78–79; economic development planning resources 79–80; economic development strategy, local food as 74–75; eligibility for 76; evaluations of grant programs 78; future effectiveness 86; impediments to evaluation of 74; institutional purchases of local food 79; management challenges for farmers 75; performance reports 78; priorities 76; purpose of 73–74; review of, need for 74; SNAP redemptions 78–79, 85; survey of market managers 80–82, *81*, **81**, **82**, 86
farmers markets, growth in 5
federal grant and loan programs 79
Feenstra, G. 52
Ferrer, M.C. 53, 54
Flora, C.B. 32–33, 36
Flora, J. 32
food deserts 18–19
food safety 13
food security, local foods and 18–19
food value chain 7–8

Garnett, T. 18
gender, demand for local food and 12
geographic hot/cold spots 9–11, *10*
Getis-Ord method 9
Gilliland, J.A. 11–12, 19
Giombolini, K.J. 55
Glanz, K. 49
Gomez, M.I. 75
Goodman, D. 7
grant programs: federal 79; *see also* Farmers Market Promotion Program (FMPP); Local Food Promotion Program (LFPP)
Green, J. 90
growth in local food systems 5, *5*

Hamm, M. 50
Hanson, J. 13
Hardesty, S.D. 49
health: food deserts 18–19; local foods and 17–18; obesity and diabetes, local food environment and 53; *see also* access to local foods, public health and
Holgate, S.T. 56–57
Holling, C.S. 90
Hughes, D.W. 15, 16, 78
human capital stocks through local foods 36–37, *38*, *39*, *41*, **42**, 44

income, demand for local food and 11–12
index of local foods **9**, 9–11, *10*
intellectual capital flows to rural areas 34–36, *35*, *36*, **39**, 39–40, 42–44, *43*
intermediaries, sales from, defining 'local' and 8
Isengildina-Massa, O. 78

Jablonski, B.B.R. 32, 65, 66
Jilcott, S.B. 53, 54
Johnson, T.G. 33
Joshi, A. 52

Kaphingst, K.M. 49
Kato, Y. 19
King, R.P. 55
Knight, A.J. 13
Kurland, N. 7

Lamie, R.D. 32
Larsen, K. 19
Lev, L. 75
Lewin, P. 33

literature on local food systems: analytical rigor as concern 16, 21–23; causality 18; characterization of local foods **9**, 9–11, *10*; community economic development 14–16; consumer demand for local foods 11–14; critiques of 21–22; defining 'local' foods 7–8, 21; early work 6; failed initiatives as overlooked 22; food deserts 18–19; food security, local foods and 18–19; future growth/potential of local foods 15; geographic hot/cold spots 9–11, *10*; 'local' assumed to be good 21–22; opportunity costs in switch to local foods 14–15; producers, impact of local foods on 16–17; public health, local foods and 17–18; quantitative index of local foods **9**, 9–11, *10*; review of 6; scalability/replicability of local foods models 16; social capital, local foods and 19–21
Local Agrifood System Sustainability/Resilience Index **93**, 94–100, **97**, *98*, **99**, 99–100
Local Food Promotion Program (LFPP): applications to 77; awards made 76; background 76–77, **77;** business plans 85; case study of grant recipients 82–83, 84–85, 86; Crossroads Community Food Network 84–85; dollar value of grants *77*; economic development strategy, local food as 74–75; evaluations of grant programs 78; future effectiveness 86; Hispanic speakers 85; immigrant entrepreneurs 85; impediments to evaluation of 74; management challenges for farmers 75; priorities 76; purpose of 73–74, 76; review of, need for 74; survey of market managers 80–82, *81*, **81**, **82**, 86; types of grants 76
local foods: defining 7–8, 21, 55; growth of systems 5, *5*; tradition of, move away from 1; *see also* access to local foods, public health and; economic development of communities; research on local food systems; resilience of local agrifood systems
Local Foods Act of 2013 (Ontario, Canada) 50
location of markets: demand for local food and 12; public health and 51–52
Low, S.A. 8
low-income countries, exports from 15
Lyson, T.A. 8

management challenges for farmers 75
Martinez, S. 55
Mayes, L. 18
McCaffrey, S.J. 7
McCutcheon, J.S. 16
McFadden, D.T. 7, 11, 16, 65
Miller, S. 78
Minaker, L.M. 53, 54
mortality, premature 54
Mowen, A.J. 58

Nie, C. 11, 12
normative tenets associated with local food 13
nutritional density of food 13

obesity, local food environment and 53
Ocloo, J.E. 57
opportunity costs in switch to local foods 14–15
organic food overlap with local production 12

Pender, J.L. 33, 37
preferences, local: demand for local food and 12; failed initiatives and 19
premature mortality 54
principal components analysis 9, 55, 56, 95, 96
producers: impact of local foods on 16–17; motivation for 16–17; social capital and 20; support programs 17
production, style of, defining 'local' and 7
promotion programs *see* Farmers Market Promotion Program (FMPP); Local Food Promotion Program (LFPP)
public health: food deserts 18–19; local foods and 17–18; obesity and diabetes, local food environment and 53; *see also* access to local foods, public health and
Purcell, M. 21, 50

quantitative index of local foods **9**, 9–11, *10*

Ragland, E. 75
Raines, N. 33
Rapid Market Assessments (RMAs) 35–36
religion(s), food and 1
replicability of local foods models 16
research on local food systems: analytical rigor as concern 16, 21–23; causality 18; characterization of local foods **9**, 9–11, *10*; community economic development 14–16; consumer demand for local foods 11–14; critiques of 21–22; defining 'local' foods 7–8, 21; early work 6; failed initiatives as overlooked 22; food deserts 18–19; food security, local foods and 18–19; future growth/potential of local foods 15; geographic hot/cold spots 9–11, *10*; opportunity costs in switch to local foods 14–15; producers, impact of local foods on 16–17; public health, local foods and 17–18; quantitative index of local foods **9**, 9–11, *10*; review of 6; scalability/replicability of local foods models 16; social capital, local foods and 19–21
resilience of local agrifood systems: comparison with other development measures 96, 99, **99;** conceptual framework 90–91; connectivity 92, **93;** county-level measurements 92; data sources 92, **93**, 99; diversity **93**, 94; ecological integrity **93**, 94; ecological resilience 90; equal weighting of indicators 95–96; Local Agrifood System Sustainability/Resilience Index

94–100, **97**, **98**, *98*, 99–100; maintenance **93**, 94; principal components analysis 95, 96; qualities necessary for 90–91, 92–94; self-organisation 93, **93;** self-organisation, need for 90–91; social capital, exclusion of general measures of 92; state-level patterns 96, **97**, *98*; usefulness of identified patterns 99
Robinson-O'Brien, R. 49
Roheim, C.A. 55
Roper, N. 78
Rossib, J. 13
rural economic development through local foods: capitals, impacts on 32–33, 37–39, **38**, **39;** Community Capitals Framework (CCF) 32; data collection 34–37, *35*, *36*; Delphi method, use of for research 33–34; evaluation of local food impact, need for 32; expert advisory panels in research 33–34; extended social accounting matrix (SAM) model 33; findings from research 37–42, *38*, *39*, *41*, **42;** government support for 31–32; human capital stocks 36–37, 40–42, *41*, **42**, 44; intellectual capital flows 34, *35*, *36*, 37, **39**, 39–40, 42–44, *43*; limited understanding of local food impacts 32; Rapid Market Assessments (RMAs) 35–36; rural areas compared to urban 31; rural wealth creation (RWC) 32; *see also* economic development of communities

Sadler, R.C. 11–12, 19
safety, food 13
Salois, M.J. 13, 53, 54
scalability of local foods models 16
Schlegel, S.A. 55
Schmit, T.M. 75
security, food 18–19
Seguin, R.A. 53
Sharp, J.S. 16
Siawor-Robertson, D. 57
Smith, M.B. 16
social capital, local foods and 19–21
socio-economic factors, demand for local food and 11–14
spatial clustering 9–11, *10*
Spoel, P. 50, 65
state support for local food markets 79
Stephenson, G. 75
Stephenson, L. 18
Stephenson, T.J. 18
Story, M. 49
Stronger Economies Together (SET) 79–80
Sundbo, D.I.C. 7
supply chains, food value chain as 7–8
sustainability of local agrifood systems: comparison with other development measures 96, 99, **99;** conceptual framework 90–91; connectivity 92, **93;** county-level measurements 92; data sources 92, **93**, 99; diversity **93**, 94; ecological integrity 94; ecological resilience 90; equal weighting of indicators 95–96; Local Agrifood System Sustainability/Resilience Index 94–100, **97**, *98*, **99**, 99–100; maintenance **93**, 94; principal components analysis 95, 96; production practices and 22; qualities necessary for 90–91, 92–94, **93;** self-organisation 93, **93**; self-organisation, need for 90–91; social capital, exclusion of general measures of 92; state-level patterns 96, **97**, *98*; usefulness of identified patterns 99
systems thinking 54

toolkit for estimating impact of local foods 80
Towe, C. 13
traceability, defining 'local' and 8
transparency, defining 'local' and 8
Tropp, D. 75
Tselepidakis, E. 12

value chain, food 7–8
value structures, food and 1
Vogel, S.J. 8

Webber, K.H. 18, 66
WIC Farmers Market Nutrition Program 52
Willis, D.B. 12
Winchester, N. 15
Woods, T. 13
Worstell, J. 90

Young, E.D. 12

Zepeda, L. 11, 12

For Product Safety Concerns and Information please contact our EU
representative GPSR@taylorandfrancis.com
Taylor & Francis Verlag GmbH, Kaufingerstraße 24, 80331 München, Germany

www.ingramcontent.com/pod-product-compliance
Lightning Source LLC
LaVergne TN
LVHW080658011026
849160LV00014B/1474
* 9 7 8 1 0 3 2 0 8 1 8 4 7 *